3

50 Leveled Math Problems

150 Problems Total

Author

Linda Dacey, Ed.D.

LESLEY
UNIVERSITY

SHELL EDUCATION

Contributing Author and Consultant

Anne M. Collins, Ph.D.
Director of Mathematics Programs
Director of Achievement Center
  for Mathematics
Lesley University

Consultants

Jayne Bamford Lynch, M.Ed.
National Faculty
Lesley University

Kerri Favreau, M.Ed.
Special Education Teacher
Cambridge Public Schools

Julie Schineller
Grade 3 Teacher
Cambridge Public Schools

Ashley Warlick, M.Ed.
Grade 3 Teacher
Cambridge Public Schools

Publishing Credits

Dona Herweck Rice, *Editor-in-Chief*; Robin Erickson, *Production Director*;
Lee Aucoin, *Creative Director*; Timothy J. Bradley, *Illustration Manager*;
Sara Johnson, M.S.Ed., *Senior Editor*; Aubrie Nielsen, M.S.Ed., *Associate Education Editor*;
Leah Quillian, *Assistant Editor*; Grace Alba, *Interior Layout Designer*; Ana Clark, *Illustrator*;
Corinne Burton, M.A.Ed., *Publisher*

Standards

© 2003 National Council of Teachers of Mathematics (NCTM)
© 2004 Mid-continent Research for Education and Learning (McREL)
© 2007 Teachers of English to Speakers of Other Languages, Inc. (TESOL)
© 2010 National Governors Association Center for Best Practices and Council of Chief State School Officers

Shell Education

5301 Oceanus Drive
Huntington Beach, CA 92649-1030
http://www.shelleducation.com
ISBN 978-1-4258-0775-7
© 2012 Shell Educational Publishing, Inc.

Table of Contents

Introduction

Problem Solving in Mathematics Instruction 5

Understanding the Problem-Solving Process. 7

Problem-Solving Strategies. 12

Ask, Don't Tell. 14

Differentiating with Leveled Problems 15

Management and Assessment. 17

How to Use This Book 22

Correlations to Standards 26

Leveled Problem-Solving Lessons

Operations and Algebraic Thinking

Snack Time 32

Floor Tiles 34

Equal Groups 36

Boxes of Cupcakes 38

Pattern Questions 40

First Names. 42

Pose a Problem. 44

Finish the Steps 46

Pattern Hunt. 48

Boxes and Boxes. 50

Figure It 52

At the Fair. 54

Yard Sale 56

What's Going On? 58

Number and Operations in Base Ten

Number Models. 60

Buildopoly. 62

Toy Store. 64

Some Sums. 66

Animal Facts. 68

Family Trips. 70

Make It True 72

Town Races. 74

Table of Contents *(cont.)*

Number and Operations—Fractions

Make It Match 76
On the Number Line 78
On the Trail. 80
Standing in Line 82
Who Is Where? 84
How Much Money? 86
What Is the Number? 88

Measurement and Data

Saturday Mornings. 90
Pose the Question 92
Balance It 94
What Does It Hold? 96
Moving Along. 98
Classroom Data 100
Measure It. 102
Keeping Track. 104
Make It Yourself 106
Which One? 108
Hobbies. 110
All Around the Garden 112
Many Measures 114

Geometry

Tangram Shapes. 116
Tell Me More 118
Name the Shape. 120
Table Shapes. 122
Shape Sentences. 124
They Belong Together 126
Draw Me 128
Parts of Shapes 130

Appendices

Appendix A: Student Response Form 132
Appendix B: Individual Observation Form 133
Appendix C: Group Observation Form 134
Appendix D: Record-Keeping Chart. 135
Appendix E: Answer Key 136
Appendix F: References Cited 141
Appendix G: Contents of the Teacher Resource CD 142

Problem Solving in Mathematics Instruction

If you were a student in elementary school before the early 1980s, your education most likely paid little or no attention to mathematical problem solving. In fact, your exposure may have been limited to solving word problems at the end of a chapter that focused on one of the four operations. After a chapter on addition, for example, you solved problems that required you to add two numbers to find the answer. You knew this was the case, so you just picked out the two numbers from the problem and added them. Sometimes, but rarely, you were assigned problems that required you to choose whether to add, subtract, multiply, or divide. Many of your teachers dreaded lessons that contained such problems as they did not know how to help the many students who struggled.

If you went to elementary school in the later 1980s or in the 1990s, it may have been different. You may have learned about a four-step model of problem solving and perhaps you were introduced to different strategies for finding solutions. There may have been a separate chapter in your textbook that focused on problem solving and two-page lessons that focused on particular problem-solving strategies, such as guess and check. Attention was given to problems that required more than one computational step for their solution, and all the information necessary to solve the problems was not necessarily contained in the problem statements.

One would think that the ability of students to solve problems would improve greatly with these changes, but that has not been the case. Research provides little evidence that teaching problem solving in this isolated manner leads to success (Cai 2010). In fact, some would argue that valuable instructional time was lost exploring problems that did not match the mathematical goals of the curriculum. An example would be learning how to use logic tables to solve a problem that involved finding out who drank which drink and wore which color shirt. Being able to use a diagram to organize information, to reason deductively, and to eliminate possibilities are all important problem-solving skills, but they should be applied to problems that are mathematically significant and interesting to students.

Today, leaders in mathematics education recommend teaching mathematics in a manner that integrates attention to concepts, skills, and mathematical reasoning. Referred to as *teaching through problem solving*, this approach suggests that problematic tasks serve as vehicles through which students acquire new mathematical concepts and skills (D'Ambrosio 2003). Students apply previous learning and gain new insights into mathematics as they wrestle with challenging tasks. This approach is quite different from introducing problems only after content has been learned.

Most recently, the *Common Core State Standards* listed the need to persevere in problem solving as the first of its Standards for Mathematical Practice (National Governors Association Center for Best Practices and Council of Chief State School Officers 2010):

Make sense of problems and persevere in solving them.

Mathematically proficient students start by explaining to themselves the meaning of a problem and looking for entry points to its solution. They analyze givens, constraints, relationships, and goals. They make conjectures about the form and meaning of the solution and plan a solution pathway rather than simply jumping into a solution attempt. They consider analogous problems, and try special cases and simpler forms of the original problem in order to

Problem Solving in Mathematics Instruction *(cont.)*

gain insight into its solution. They monitor and evaluate their progress and change course if necessary. Older students might, depending on the context of the problem, transform algebraic expressions or change the viewing window on their graphing calculator to get the information they need. Mathematically proficient students can explain correspondences between equations, verbal descriptions, tables, and graphs or draw diagrams of important features and relationships, graph data, and search for regularity or trends. Younger students might rely on using concrete objects or pictures to help conceptualize and solve a problem. Mathematically proficient students check their answers to problems using a different method, and they continually ask themselves, "Does this make sense?" They can understand the approaches of others to solving complex problems and identify correspondences between different approaches.

This sustained commitment to problem solving makes sense; it is the application of mathematical skills to real-life problems that makes learning mathematics so important. Unfortunately, we have not yet mastered the art of developing successful problem solvers. Students' performance in the United States on the 2009 Program for International Student Assessment (PISA), a test that evaluates 15-year-old students' mathematical literacy and ability to apply mathematics to real-life situations, suggests that we need to continue to improve our teaching of mathematical problem solving. According to data released late in 2010, students in the U.S. are below average (National Center for Educational Statistics 2010). Clearly we need to address this lack of success.

Students do not have enough opportunities to solve challenging problems. Further, problems available to teachers are not designed to meet the individual needs of students. Additionally, teachers have few opportunities to learn how best to create, identify, and orchestrate problem-solving tasks. *50 Leveled Math Problems* is a unique series that is designed to address these concerns.

Understanding the Problem-Solving Process

George Polya is known as the father of problem solving. In his book *How to Solve It: A New Aspect of Mathematical Method* (1945), Polya provides a four-step model of problem solving that has been adopted in many classrooms: understanding the problem, making a plan, carrying out the plan, and looking back. In some elementary classrooms this model has been shortened to: understand, plan, do, check. Unfortunately, this over-simplification ignores much of the richness of Polya's thinking.

Polya's conceptual model of the problem-solving process has been adapted for use at this level. Teachers are encouraged to view the four steps as interrelated, rather than only sequential, and to recognize that problem-solving strategies are useful at each stage of the problem-solving process. The model presented here gives greater emphasis to the importance of communicating and justifying one's thinking as well as to posing problems. Ways in which understanding is deepened throughout the problem-solving process is considered in each of the following steps.

Step 1: Understand the Problem

Students engage in the problem-solving process when they attempt to *understand the problem*, but the understanding is not something that just happens in the beginning. At grade 3, students may be asked to restate the problem in their own words and then turn to a partner to summarize what they know and what they need to find out. The teacher may read the problem aloud when working with a small group of English language learners.

What is most important is that teachers do not teach students to rely on key words or show students "tricks" or "short-cuts" that are not built on conceptual understanding. Interpreting the language of mathematics is complex, and terms that are used in mathematics often have different everyday meanings. Note how a reliance on key words would lead to failure when solving the problem below. A student taught that *total* means *add* may decide that 6 + 3 + 4, or 13 pens, is the correct answer:

> *Nathan bought some pens.*
> *He gave 6 pens to each of his 3 daughters.*
> *He kept 4 pens for himself.*
> *What is the total number of pens Nathan bought?*

Step 2: Apply Strategies

Once students have a sense of the problem they can begin to actively explore it. They may do so by applying one or more of the strategies below. Note that related actions are combined within some of the strategies.

- Act it out or use manipulatives.
- Count, compute, or write an equation.
- Find information in a picture, list, table, graph, or diagram.
- Generalize a pattern.
- Guess and check or make an estimate.
- Organize information in a picture, list, table, graph, or diagram.
- Simplify the problem.
- Use logical reasoning.
- Work backward.

Understanding the Problem-Solving Process (cont.)

Step 2: Apply Strategies (cont.)

As students apply these strategies, they also deepen their understanding of the mathematics of the problem. As such, understanding develops throughout the problem-solving phases. Consider the following problem and scenario in which students use each number from the box once in the blanks to make the equations true.

57	195	147	257	746	108

$$342 \ = \ \rule{2cm}{0.4pt} \ + \ \rule{2cm}{0.4pt}$$

$$\rule{2cm}{0.4pt} \ - \ 489 \ = \ \rule{2cm}{0.4pt}$$

$$\rule{2cm}{0.4pt} \ + \ 51 \ = \ \rule{2cm}{0.4pt}$$

Nick and Lauren are working together. They have read the problem and understand that they are to write the numbers so that the equations are true. Lauren sighs and says, "We'll never get this. There are too many choices." Nick responds, "I agree, but maybe we could try something. Let's just write the numbers and see what happens." Lauren agrees and proceeds to fill in the numbers randomly. As she does so, she says, "Wait, 746 can't go here; the whole thing is only 342." She continues, "And it can't be added to 51 either because it is the biggest number." Nick then builds on her thinking, "So it has to go before 489. We are totally getting this."

By making a guess and checking it, Nick and Lauren came to understand necessary relationships among the numbers they chose and the other numbers in the equation.

It is important that we offer students problems that can be solved in more than one way. If one strategy does not lead to success, students can try a different one. This option gives students the opportunity to learn that getting "stuck" might just mean that a new approach should be considered. When students get themselves "unstuck" they are more likely to view themselves as successful problem solvers. Such problems also lead to richer mathematical conversations as there are different ideas and perspectives to discuss. Consider the following problem:

> *There are 4 baskets. There are 5 apples in each basket.*
> *Jamie takes an apple from each basket to give to her friends.*
> *In all, how many apples are left?*

As the following samples of student responses indicate, organizing information in a drawing, computing, and writing an equation are common strategies, though they are used in different ways. Student Sample 1 suggests how much detail some students prefer to include in their drawings. We see baskets, apples, and friends. Note the balance in the way the student has shown the apples being taken from the baskets and given to the friends. An addition sentence has also been included.

Understanding the Problem-Solving Process *(cont.)*

Step 2: Apply Strategies *(cont.)*

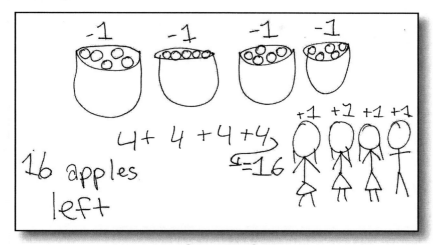

Student Sample 1

Some students may also use addition, but not be inclined to make a drawing to represent the situation. In Student Sample 2, the student finds the total number of apples first, and then subtracts the four apples given to friends. As such, the student continues the equation instead of starting a new one. This student should be encouraged to record his response in two equations: 10 + 10 = 20 and 20 − 4 = 16.

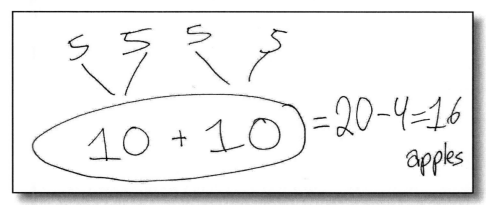

Student Sample 2

Student Sample 3 on the next page combines many aspects of both of the students' thinking from Samples 1 and 2. Although more simple, a drawing is included that indicates the original groups of five apples and the one apple that was taken from each group. This student records two equations—one that represents the groups once the single apples have been removed, and one that finds the total number and then subtracts the four apples.

Understanding the Problem-Solving Process *(cont.)*

Step 2: Apply Strategies *(cont.)*

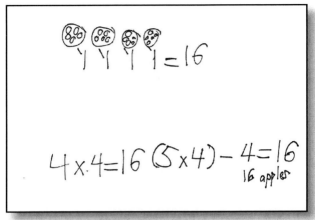

Student Sample 3

Step 3: Communicate and Justify Your Thinking

It is essential that we ask students to communicate and justify their thinking. It is also important that students make records as they work so that they can recall their thinking. When we make it clear that we expect such behavior from students, we are establishing an important habit of mind (Goldenberg, Shteingold, and Feurzeig 2003) and developing their understanding of the nature of mathematics. We may also discover potential misconceptions. Consider the following problem:

> *Philip and Rosa are skiing down a ski trail.*
> *Philip is $\frac{2}{6}$ of the way down the trail.*
> *Rosa is $\frac{2}{8}$ of the way down the trail.*
> *Who is farther down the trail?*

As shown in the response below, a student can identify the correct answer *(Philip)*, without necessarily understanding the importance of the relationship between the numerator and denominator. If this teacher had merely evaluated a correct answer, she might not have known to ask the follow-up question, *What if Rosa was $\frac{4}{8}$ of the way down the trail?*

Phillip is farther down the trial because his denominator is the lowest and the lower it is the bigger amount.

Student Sample 4

Understanding the Problem-Solving Process *(cont.)*

Step 3: Communicate and Justify Your Thinking *(cont.)*

When students explain their thinking orally while investigating a problem with a partner or small group, they may deepen their understanding of the problem or recognize an error and fix it. When students debrief after finding solutions, they learn to communicate their thinking clearly and in ways that give others access to new mathematical ideas. In one third grade class, a student observes, "It's better when you estimate first. You can save yourself from having to check a lot of guesses." Such discourse is essential to the mathematical practice suggested in the *Common Core State Standards* that students "construct viable justifications and critique the reasoning of others" (National Governors Association Center for Best Practices and Council of Chief State School Officers 2010).

Our task is to foster learning environments where students engage in this kind of *accountable talk*. Michaels, O'Connor, and Resnick (2008) identify three aspects of this type of dialogue. The first is that students are accountable to their learning communities; they listen to each other carefully and build on the ideas of others. Second, accountable talk is based on logical thinking and leads to logical conclusions. Finally, these types of discussions are based on facts or other information that is available to everyone.

When we emphasize the importance of discussions and explanations, we are teaching our students that it is the soundness of their mathematical reasoning that determines what is correct, not merely an answer key or a teacher's approval. Students learn, therefore, that mathematics makes sense and that they are mathematical sense-makers.

Step 4: Take It Further

Debrief

It is this final step in the problem-solving process to which teachers and students are most likely to give the least attention. When time is given to this step, it is often limited to "check your work." In contrast, this step offers rich opportunities for further learning. Students might be asked to solve the problem using a different strategy, or to find additional solutions. They might be asked to make a mathematical generalization based on their investigation. Students might connect this problem to another problem they have solved already, or they now may be able to solve a new, higher-level problem.

Posing Problems

Students can also take problem solving further by posing problems. In fact, problem posing is intricately linked with problem solving (Brown and Walter 2005). When posing their own problems, students can view a problem as something they can create, rather than as a task that is given to them. This book supports problem posing through a variety of formats. For example, students may be asked to supply missing data in a problem so that it makes sense. They may be given a problem with the question omitted and asked to compose one. Or, they may be given both problem data and the answer and asked to identify the missing question. Teachers may also choose to ask students to create their own problems that are similar to those they have previously solved. Emphasis on problem posing can transform the teaching of problem solving and build lifelong curiosity in students.

Problem-Solving Strategies

Think of someone doing repair jobs around the house. Often that person carries a toolbox or wears a tool belt from task to task. Common tools such as hammers, screwdrivers, and wrenches are then readily available. The repair person chooses tools (usually more than one) appropriate for a particular task. Problem-solving strategies are the tools used to solve problems. Labeling the strategies allows students to refer to them in discussions and helps students recognize the wide variety of tools available for the solution of problems. The problems in this book provide opportunities for students to apply one or more of the following strategies:

Act It Out or Use Manipulatives

Students' understanding of a problem is greatly enhanced when they act it out. Students may choose to dramatize a situation themselves or use manipulatives to show the actions or changes that take place. If students suggest they do not understand a problem say something such as *Imagine this is a play. Show me what the actors are doing.*

Count, Compute, or Write an Equation

When students count, compute, or write an equation to solve a problem they are making a match between a context and a mathematical skill. Once the connection is made, students need only to carry out the procedure accurately. Sometimes writing an equation is a final step in the solution process. For example, students might work with manipulatives or draw pictures and then summarize their thinking by recording an equation.

Find Information in a Picture, List, Table, Graph, or Diagram

Too often problems contain all of the necessary information in the problem statement. Such information is never so readily available in real-world situations. It is important that students develop the ability to interpret a picture, list, table, graph, or diagram and identify the information relevant to the problem.

Generalize a Pattern

Some people consider mathematics the study of patterns, so it makes sense that the ability to identify, continue, and generalize patterns is an important problem-solving strategy. The ability to generalize a pattern requires students to recognize and express relationships. Once generalized, the student can use the pattern to predict other outcomes.

Guess and Check or Make an Estimate

Guessing and checking or making an estimate provide students with insights into problems. Making a guess can help students to better understand conditions of the problem; it can be a way to try something when a student is stuck. Some students may make random guesses, but over time, students learn to make more informed guesses. For example, if a guess leads to an answer that is too large, a student might next try a number that is less than the previous guess. Estimation can help students narrow their range of guesses or be used to check a guess.

Problem-Solving Strategies (cont.)

Organize Information in a Picture, List, Table, Graph, or Diagram

Organizing information can help students both understand and solve problems. For example, students might draw a number line or a map to note information given in the problem statement. When students organize data in a table or graph they might recognize relationships among the data. Students might also make an organized list to keep track of guesses they have made or to identify patterns. It is important that students gather data from a problem and organize it in a way that makes the most sense to them.

Simplify the Problem

Another way for students to better understand a problem, or perhaps get "unstuck," is to simplify it. Often the easiest way for students to do this is to make the numbers easier. For example, a student might replace four-digit numbers with single-digit numbers or replace fractions with whole numbers. With simpler numbers students often gain insights or recognize relationships that were not previously apparent, but that can now be applied to the original problem. Students might also work with 10 numbers, rather than 100, to identify patterns.

Use Logical Reasoning

Logical thinking and sense-making pervade mathematical problem solving. To solve problems students need to deduce relationships, draw conclusions, make inferences, and eliminate possibilities. Logical reasoning is also a component of many other strategies. For example, students use logical reasoning to revise initial guesses or to interpret diagrams. Asking questions such as *What else does this sentence tell you?* helps students to more closely analyze given data.

Work Backward

When the outcome of a situation is known, we often work backward to determine how to arrive at that goal. We might use this strategy to figure out what time to leave for the airport when we know the time our flight is scheduled to depart. A student might work backward to answer the question *What did Joey add to 79 to get a sum of 146?* or *If it took 2 hours and 23 minutes to drive a given route and the driver arrived at 10:17, at what time did the driver leave home?* Understanding relationships among the operations is critical to the successful use of this strategy.

Ask, Don't Tell

All teachers want their students to succeed, and it can be difficult to watch them struggle. Often when students struggle with a problem, a first instinct may be to step in and show them how to solve it. That intervention might feel good, but it is not helpful to the student. Students need to learn how to struggle through the problem-solving process if they are to enhance their understanding and reasoning skills. Perseverance in solving problems is listed under the mathematical practices in the *Common Core State Standards* and research indicates that students who struggle and persevere in solving problems are more likely to internalize the problem-solving process and build upon their successes. It is also important to recognize the fact that people think differently about how to approach and solve problems.

An effective substitution for telling or showing students how to solve problems is to offer support through questioning. George Bright and Jeane Joyner (2005) identify three different types of questions to ask, depending on where students are in the problem-solving process: (1) engaging questions, (2) refocusing questions, and (3) clarifying questions.

Engaging Questions

Engaging questions are designed to pique student interest in a problem. Students are more likely to want to solve problems that are interesting and relevant. One way to immediately grab a student's attention is by using his or her name in the problem. Once a personal connection is made, a student is more apt to persevere in solving the problem. Posing an engaging question is also a great way to redirect a student who is not involved in a group discussion. Suppose students are provided the missing numbers in a problem and one of the sentences reads *Janel is about _____ centimeters tall and rides her bicycle to school.* Engaging questions might include *What do you know about 100 centimeters? Are you taller or shorter than 100 centimeters?* The responses will provide further insight into how the student is thinking.

Refocusing Questions

Refocusing questions are asked to redirect students away from a nonproductive line of thinking and back to a more appropriate track. These questions often begin with the phrase *What can you tell me about…?* or *What does this number…?* Refocusing questions are also appropriate if you suspect students have misread or misunderstood the problem. Asking them to explain in their own words what the problem is stating and what question they are trying to answer is often helpful.

Clarifying Questions

Clarifying questions are posed when it is unclear why students have used a certain strategy, picture, table, graph, or computation. They are designed to help demonstrate what students are thinking, but can also be used to clear up misconceptions students might have. The teacher might say *I am not sure why you started with the number 10. Can you explain that to me?*

As teachers transform instruction from "teaching as telling" to "teaching as facilitating," students may require an adjustment period to become accustomed to the change in expectations. Over time, students will learn to take more responsibility and to expect the teacher to probe their thinking, rather than supply them with answers. After making this transition in her own teaching, one teacher shared a student's comment: "I know when I ask you a question that you are only going to ask me a question in response. But, sometimes the question helps me figure out the next step I need to take. I like that."

Differentiating with Leveled Problems

There are four main ways that teachers can differentiate: by content, by process, by product, and by learning environment. Differentiation by content involves varying the material that is presented to students. Differentiation by process occurs when a teacher delivers instruction to students in different ways. Differentiation by product asks students to present their work in different ways. Offering different learning environments, such as small group settings, is another method of differentiation. Students' learning styles, readiness levels, and interests determine which differentiation strategies are implemented. The leveled problems in this book vary aspects of mathematics problems so that students at various readiness levels can succeed. Mini-lessons include problems at three levels and ideas for differentiation. These are designated by the following symbols:

- ● lower-level challenge

- ■ on-level challenge

- ▲ above-level challenge

- ★ English language learner support

Ideally, students solve problems that are at just the right level of challenge—beyond what would be too easy, but not so difficult as to cause extreme frustration (Sylwester 2003; Tomlinson 2003; Vygotsky 1986). The goal is to avoid both a lack of challenge, which might leave students bored, as well as too much of a challenge, which might lead to significant anxiety.

There are a variety of ways to level problems. In this book, problems are leveled based on the concepts and skills required to find the solution. Problems are leveled by adjusting one or more of the following factors:

Complexity of the Mathematical Language

The mathematical language used in problems can have a significant impact on their level of challenge. For example, negative statements are more difficult to interpret than positive ones. So, *It is not an even number* is more complex than *It is an odd number.* Phrases such as *at least* or *between* also add to the complexity of the information. Further, words such as *table, face,* and *plot* can be challenging since their mathematical meaning differs from their everyday uses.

Complexity of the Task

There are various ways to change the complexity of the task. One example would be the number of solutions that students are expected to identify. Finding one solution that satisfies problem conditions is less challenging than finding more than one solution, which is even less difficult than identifying *all* possible solutions. Similarly, increases and decreases in the number of conditions that must be met and the number of steps that must be completed change the complexity of a problem.

Differentiating with Leveled Problems (cont.)

Changing the Numbers

Sometimes it is the size of the numbers that is changed to increase the level of mathematical skill required. A problem may be more complex when it involves three-digit numbers rather than two-digit numbers. Sometimes changes to the "friendliness" of the numbers are made to adapt the difficulty level. For example, two problems may involve basic facts, but students are likely to find one that involves five as a factor easier than one that involves seven as a factor.

Amount of Support

Some problems provide more support for learners than others. Providing a graphic organizer or a table that is partially completed is one way to provide added support for students. Offering information with pictures rather than words can also vary the level of support. The inclusion of such supports often helps students to better understand problems and may offer insights on how to proceed. The exclusion of supports allows a learner to take more responsibility for finding a solution, and it may make the task appear more abstract or challenging.

Differentiation Strategies for English Language Learners

Many English language learners may work at a high readiness level in many mathematical concepts, but may need support in accessing the language content. Specific suggestions for differentiating for English language learners can be found in the *Differentiate* section of some of the mini-lessons. Additionally, the strategies below may assist teachers in differentiating for English language learners.

- Allow students to draw pictures or provide oral responses as an alternative to written responses.
- Pose questions with question stems or frames. Example question stems/frames include:
  - *What would happen if…?*
  - *Why do you think…?*
  - *How would you prove…?*
  - *How is _____ related to _____?*
  - *Why is _____ important?*
- Use visuals to give context to questions. Add pictures or icons next to key words, or use realia to help students understand the scenario of the problem.
- Provide sentence stems or frames to help students articulate their thoughts. Sentence stems include:
  - *This is important because…*
  - *This is similar because…*
  - *This is different because….*

  Sentence frames include:
  - *I agree with _____ because…*
  - *I disagree with _____ because…*
  - *I think _____ because….*
- Partner English language learners with language-proficient students.

Management and Assessment

Organization of the Mini-Lessons

The mini-lessons in this book are organized according to the domains identified in the *Common Core State Standards*, which have also been endorsed by the National Council of Teachers of Mathematics. At grade 3, these domains are *Operations and Algebraic Thinking, Number and Operations in Base Ten, Number and Operations—Fractions, Measurement and Data*, and *Geometry*. Though organized in this manner, the mini-lessons are independent of one another and may be taught in any order within a domain or among the domains. What is most important is that the lessons are implemented in the order that best fits a teacher's curriculum and practice.

Ways to Use the Mini-Lessons

There are a variety of ways to assign and use the mini-lessons, and they may be implemented in different ways throughout the year. The lessons can provide practice with new concepts or be used to maintain skills previously learned. The problems can be incorporated into a teacher's mathematics lessons once or twice each week, or they may be used to introduce extended or additional instructional periods. They can be used in the regular classroom with the whole class or in small groups. They can also be used to support Response to Intervention (RTI) and after-school programs.

It is important to remember that a student's ability to solve problems depends greatly on the specific content involved and may change over the course of the school year. Establish the expectation that problem assignment is flexible; sometimes students will be assigned to one level (circle, square, or triangle) and sometimes to another. On occasion, you may also wish to allow students to choose their own problems. Much can be learned from students' choices!

Students can also be assigned one, two, or all three of the problems to solve. Although leveled, some students who are capable of wrestling with complex problems need the opportunity to warm up first to build their confidence. Starting at a lower level serves these students well. Teachers may also find that students correctly assigned to a below- or on-level problem will be able to consider a problem at a higher level after solving one of the lower problems. Students can also revisit these problems, investigating those at the higher levels not previously explored.

Julie Schineller, a third-grade teacher, reflects on the results of allowing her students to choose which leveled problem to complete:

> *Generally when I differentiate in my class I believe that I know the levels of my students and give them problems that best match their mathematical abilities. So, I control which problems the students do. When I saw these leveled problems, I wanted to try something different.*
>
> *I have been working hard over the last few years to hold my students more accountable for their own learning and goal-setting. I gave everyone all three levels of the problems and had them consider which problem they thought would be "just right" for them. We compared it to literacy when they pick a book that is not too easy or too hard, but will help them to grow and meet their desired goals.*

Management and Assessment *(cont.)*

Ways to Use the Mini-Lessons *(cont.)*

The feedback was overwhelming. They loved having the opportunity to make these decisions for themselves. I anticipated that the kids who love a challenge would go for the harder problems, and for the most part that was true. However, one of my students shared that she started with the easier problems. She said, "My parents told me when I have a lot to do I should do the easier things first and save the hardest things for last so that I can try my hardest for a long time."

Several students shared that for each set of problems they would read all of the problems and then decide what kind of math the problems involved. For example, one student said, "So for these problems I could do either multiplication or division, and I am really good at that. I am going to start with the hardest problem for these." Another student offered, "I'm not very good with fractions. So I am going to do the middle one—not too hard, not too easy—and see how I do."

I learned a lot about my students by giving them this choice. I was impressed by how well they could articulate their strengths and weaknesses. The students made appropriate decisions and often surprised me by also trying the problem at the higher level.

Grouping Students to Solve Leveled Problems

A differentiated classroom often groups students in a variety of ways, based on the instructional goals of an activity or the tasks students must complete. At times, students may work in heterogeneous groups or pairs of students of varying readiness levels. Other activities may lend themselves to homogeneous groups or pairs of students who share similar readiness levels. Since the problems presented in this book provide below-level, on-level, and above-level challenges, you may wish to partner or group students with others who are working at the same readiness levels.

Since students' readiness levels may vary for different mathematical concepts and change throughout a course of study, students may be assigned different levels of problems at different times throughout the year (or even throughout a week). It is important that the grouping of students for solving leveled problems stay flexible. Struggling students who feel that they are constantly assigned to work with a certain partner or group may develop feelings of shame or stigma. Above-level students who are routinely assigned to the same group may become disinterested and cause behavior problems. Varying students' groups can help keep the activities interesting and engaging.

Management and Assessment *(cont.)*

Assessment for Learning

In recent years, increased attention has been given to summative assessment in schools. Significantly more instructional time is taken with weekly quizzes, chapter tests, and state-mandated assessments. These tests, although seen as tedious by many, provide information and reports about achievement to students, parents, administrators, and other interested stakeholders. However, these summative assessments often do not have a real impact on an individual student's learning. In fact, when teachers return quizzes and tests, many students look at the grade and if it is "good," they bring the assessment home. If it is not an acceptable grade, they often just throw away the assessment.

Research shows that to have an impact on student learning we should rely on assessments *for* learning, rather than assessments *of* learning. That is, we should focus on assessment data we collect during the learning process, not after the instructional cycle is completed. These assessments for learning, or formative assessments, are shown to have the greatest positive impact on student achievement (National Mathematics Advisory Panel 2008). Assessment for learning is an ongoing process that includes a variety of strategies and protocols to inform the progression of student learning.

One might ask, "So, what is the big difference? Don't all assessments accomplish the same goal?" The answer to those key questions is *no*. A great difference is the fact that formative assessment is designed to make student thinking visible. This is a real transformation for many teachers because when the emphasis is on student thinking and reasoning, the focus shifts from whether the answer is correct or incorrect to how the students grapple with a problem. Making student thinking visible entails a change in the manner in which teachers interact with their students. For instance, instead of relying solely on students' written work, teachers gather information through observation, questioning, and listening to their students discuss strategies, justify their reasoning, and explain why they chose to make particular decisions or use a specific representation. Since observations happen in real time, teachers can react in the moment by making an appropriate instructional decision, which may mean asking a well-posed question or suggesting a different model to represent the problem at hand.

Students are often asked to explain what they were thinking as they completed a procedure. Their response is often a recitation of the steps that were used. Such an explanation does not shed any light on whether a student understands the procedure, why it works, or if it will always work. Nor does it provide teachers with any insight into whether a student has a superficial or a deep understanding of the mathematics involved. If, however, students are encouraged to explain their thought processes, teachers will be able to discern the level of understanding. The vocabulary students use (or do not use) and the confidence with which they are able to answer probing questions can also provide insight into their levels of comprehension.

One of the most important features of formative assessment is that it actively involves students in their own learning. In assessment for learning, students are asked to reflect on their own work. They may be asked to consider multiple representations of a problem and then decide which of those representations makes the most sense, or which is the most efficient, or how they relate to one another. Students may be asked to make conjectures and then prove or disprove them by negation or counterexamples. Notice that it is the students doing the hard work of making decisions and thinking through the mathematical processes. Students who work at this level of mathematics, regardless of their grade level, demonstrate a deep understanding of mathematical concepts.

Management and Assessment (cont.)

Assessment for Learning (cont.)

Assessment for learning makes learning a shared endeavor between teachers and students. In effective learning environments, students take responsibility for their learning and feel safe taking risks, and teachers have opportunities to gain a deeper understanding of what their students know and are able to do. Implementing a variety of tools and protocols when assessing for learning can help the process become seamless. Some specific formative assessment tools and protocols include:

- Student Response Forms or Journals
- Range Questions
- Gallery Walks
- Observation Protocols
- Feedback
- Exit Cards

Student Response Forms or Journals

Providing students with an organized workspace for the problems they solve can help a teacher to better understand a student's thinking and more easily identify misconceptions. Students often think that recording an answer is enough. If students do include further details, they often only write enough to fill the limited space that might be provided on an activity sheet. To promote the expectation that students show all of their work and record more of their thinking, use the included *Student Response Form* (page 132; studentresponse.pdf), or have students use a designated journal or notebook for solving problems. The prompts on the *Student Response Form* and the additional space provided encourage students to offer more details.

Range Questions

Range questions allow for a variety of responses and teachers can use them to quickly gain access to students' understanding. Range questions are included in the activate section of many mini-lessons. The questions or problems that are posed are designed to provide insight into the spectrum of understanding that your students bring to the day's problems. For instance, you might ask *What does it mean to have groups that are equal in number? When might we want to be sure that groups have the same number of people?* As you can imagine, the level of sophistication in the responses would vary and can help you decide which students to assign to which of the leveled problems.

Gallery Walks

Gallery walks can be used in many ways, but they all promote the sharing of students' problem-solving strategies and solutions. Pairs or small groups of students can record their pictures, tables, graphs, diagrams, computational procedures, and justifications on chart paper that they hang in designated areas of the classroom prior to the debriefing component of the lesson. Or, simply have students place their *Student Response Forms* at their workspaces and have students take a tour of their classmates' thinking. Though suggested occasionally for specific mini-lessons, you can include this strategy with any of the mini-lessons.

Management and Assessment *(cont.)*

Assessment for Learning *(cont.)*

Observation Protocols

Observation protocols facilitate the data gathering that teachers must do as they document evidence of student learning. Assessment of learning is a key component in a teacher's ability to say, "I know that my students can apply these mathematical ideas because I have this evidence." Some important learning behaviors for teachers to focus on include: level of engagement in the problem/task; incorporation of multiple representations; inclusion of appropriate labels in pictures, tables, graphs, and solutions; use of accountable talk; inclusion of reflection on their work; and connections made between and among other mathematical ideas, previous problems, and their own life experiences. There is no one right form, nor could all of these areas be included on a form while leaving room for comments. Protocols should be flexible and allow you to identify categories of learning important to a teacher and his or her students. Two observation forms are provided in the appendices—one can be used with individual students (page 133; individualobs.pdf), and one can be used when observing a group (page 134; groupobs.pdf).

Feedback

Feedback is a critical component of formative assessment. Teachers who do not give letter grades on projects, quizzes, or tests, but who provide either neutral feedback or inquisitive feedback, find their students take a greater interest in the work they receive back than they did when their papers were graded. There are different types of feedback, but effective feedback focuses on the evidence in student work. Many students respond favorably to an "assessment sandwich." The first comment might be a positive comment or praise for something well done, followed by a critical question or request for further clarification, followed by another neutral or positive comment.

Exit Cards

Exit cards are an effective way of assessing students' thinking at the end of a lesson in preparation for future instruction. There are multiple ways in which exit cards can be used. A similar problem to the one students have previously solved can be posed, or students can be asked to identify topics of confusion, what they liked best, or what they think they learned from a lesson. One simple exit-card task involves providing students with two sentence starters: *I learned that…* and *I need more practice with….* A template for this exit card is provided on the Teacher Resource CD (exitcard.pdf). Exit-card tasks are suggested in the *Differentiate* sections of some of the mini-lessons, but they may be added to any mini-lesson.

How to Use This Book

Mini-Lesson Plan

Lessons are organized by **Common Core State Standard** domains.

Suggested **Problem-Solving Strategies** outline strategies students may want to use in solving the problem. However, these are not the only strategies that can be used to solve the problem.

The McREL mathematics **Standards** for each lesson are provided.

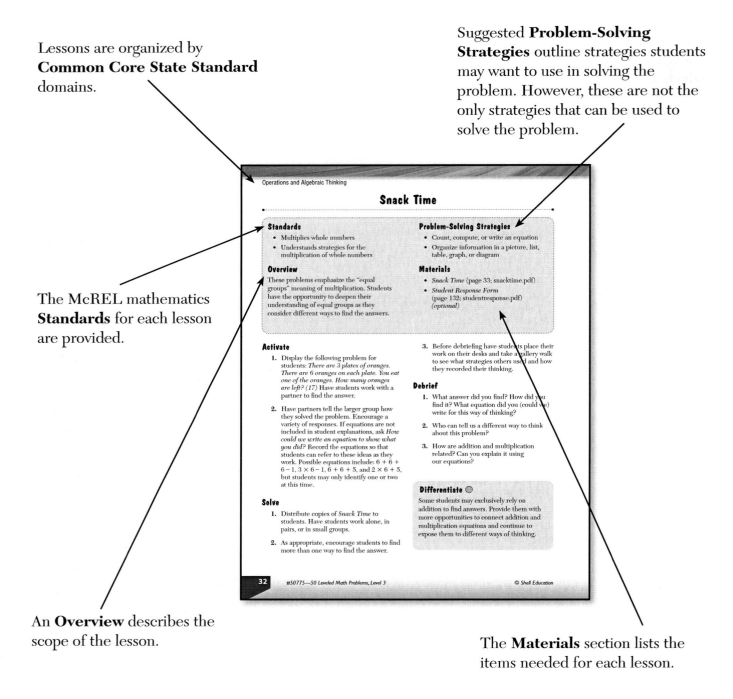

Operations and Algebraic Thinking

Snack Time

Standards
- Multiplies whole numbers
- Understands strategies for the multiplication of whole numbers

Overview
These problems emphasize the "equal groups" meaning of multiplication. Students have the opportunity to deepen their understanding of equal groups as they consider different ways to find the answers.

Problem-Solving Strategies
- Count, compute, or write an equation
- Organize information in a picture, list, table, graph, or diagram

Materials
- *Snack Time* (page 33; snacktime.pdf)
- *Student Response Form* (page 132; studentresponse.pdf) *(optional)*

Activate
1. Display the following problem for students: *There are 3 plates of oranges. There are 6 oranges on each plate. You eat one of the oranges. How many oranges are left?* (17) Have students work with a partner to find the answer.
2. Have partners tell the larger group how they solved the problem. Encourage a variety of responses. If equations are not included in student explanations, ask *How could we write an equation to show what you did?* Record the equations so that students can refer to these ideas as they work. Possible equations include: $6 + 6 + 6 - 1$, $3 \times 6 - 1$, $6 + 6 + 5$, and $2 \times 6 + 5$, but students may only identify one or two at this time.

Solve
1. Distribute copies of *Snack Time* to students. Have students work alone, in pairs, or in small groups.
2. As appropriate, encourage students to find more than one way to find the answer.

3. Before debriefing have students place their work on their desks and take a gallery walk to see what strategies others used and how they recorded their thinking.

Debrief
1. What answer did you find? How did you find it? What equation did you (could we) write for this way of thinking?
2. Who can tell us a different way to think about this problem?
3. How are addition and multiplication related? Can you explain it using our equations?

Differentiate ◯
Some students may exclusively rely on addition to find answers. Provide them with more opportunities to connect addition and multiplication equations and continue to expose them to different ways of thinking.

32 #50775—50 Leveled Math Problems, Level 3   © Shell Education

An **Overview** describes the scope of the lesson.

The **Materials** section lists the items needed for each lesson.

How to Use This Book *(cont.)*

Mini-Lesson Plan *(cont.)*

The **Activate** section suggests how you can access or assess students' prior knowledge. This section might recommend ways to have students review vocabulary, recall experiences related to the problem contexts, remember relevant mathematical ideas, or solve simpler related problems.

The **Solve** section provides suggestions on how to group students for the problem they will solve. It also provides questions to ask, observations to make, or procedures to follow to guide students in their work.

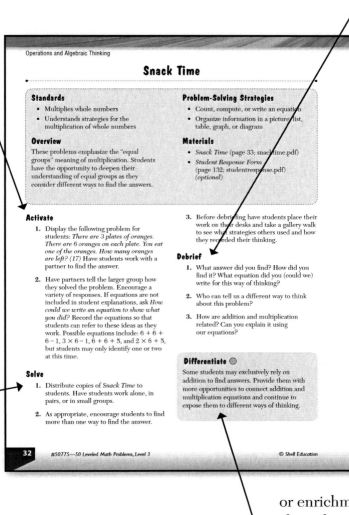

Operations and Algebraic Thinking

Snack Time

Standards
- Multiplies whole numbers
- Understands strategies for the multiplication of whole numbers

Overview
These problems emphasize the "equal groups" meaning of multiplication. Students have the opportunity to deepen their understanding of equal groups as they consider different ways to find the answers.

Problem-Solving Strategies
- Count, compute, or write an equation
- Organize information in a picture, list, table, graph, or diagram

Materials
- *Snack Time* (page 33; snacktime.pdf)
- *Student Response Form* (page 132; studentresponse.pdf) *(optional)*

Activate
1. Display the following problem for students: *There are 3 plates of oranges. There are 6 oranges on each plate. You eat one of the oranges. How many oranges are left?* (17) Have students work with a partner to find the answer.
2. Have partners tell the larger group how they solved the problem. Encourage a variety of responses. If equations are not included in student explanations, ask *How could we write an equation to show what you did?* Record the equations so that students can refer to these ideas as they work. Possible equations include: $6 + 6 + 6 - 1$, $3 \times 6 - 1$, $6 + 6 + 5$, and $2 \times 6 + 5$, but students may only identify one or two at this time.

Solve
1. Distribute copies of *Snack Time* to students. Have students work alone, in pairs, or in small groups.
2. As appropriate, encourage students to find more than one way to find the answer.

3. Before debriefing have students place their work on their desks and take a gallery walk to see what strategies others used and how they recorded their thinking.

Debrief
1. What answer did you find? How did you find it? What equation did you (could we) write for this way of thinking?
2. Who can tell us a different way to think about this problem?
3. How are addition and multiplication related? Can you explain it using our equations?

Differentiate ⬤
Some students may exclusively rely on addition to find answers. Provide them with more opportunities to connect addition and multiplication equations and continue to expose them to different ways of thinking.

32 #50775—50 Leveled Math Problems, Level 3   © Shell Education

The **Debrief** section provides questions designed to deepen students' understanding of the mathematics and the problem-solving process. Because the leveled problems share common features, it is possible to debrief either with small groups or as a whole class.

The **Differentiate** section includes additional suggestions to meet the unique needs of students. This section may offer support for English language learners, scaffolding for below-level students, or enrichment opportunities for above-level students. The following symbols are used to indicate appropriate readiness levels for the provided differentiation:

⬤   below level

▢   on level

△   above level

☆   English language learner

How to Use This Book (cont.)

Lesson Resources

Leveled Problems

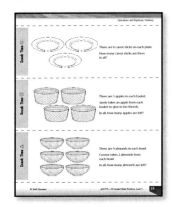

Each activity sheet offers **leveled problems** at three levels of challenge—below level, on level, and above level. Cut the activity sheet apart and distribute the appropriate problem to each student, or present all of the leveled problems on an activity sheet to every student.

Record-Keeping Chart

Use the **Record-Keeping Chart** (page 135) to keep track of the problems each student completes.

Observation Forms

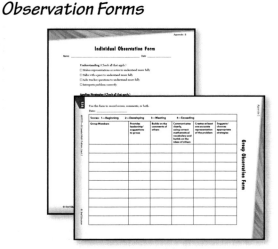

Use the **Individual Observation Form** (page 133) to document students' progress as they work through problems on their own. Use the **Group Observation Form** (page 134) to keep a record of students' success in working with their peers to solve problems.

Teacher Resource CD

Helpful reproducibles are provided on the accompanying **Teacher Resource CD**. A detailed listing of the CD contents can be found on pages 142–144. The CD includes:

- Resources to support the implementation of the mini-lessons
- Manipulative templates
- Reproducible PDFs of all leveled problems and assessment tools
- Correlations to standards

How to Use This Book (cont.)

Lesson Resources (cont.)

Student Response Form

Students can attach their leveled problem to the form.

Students have space to show their work, provide their solution, and explain their thinking.

Appendix A

Name: _____   Date: _____

Student Response Form

Problem:

(glue your problem here)

My Work and Illustrations:
(picture, table, list, graph)

My Solution:

My Explanation:

_____

_____

_____

132   #50775—50 Leveled Math Problems, Level 3   © Shell Education

---

Correlations to Standards

Shell Education is committed to producing educational materials that are research- and standards-based. In this effort, we have correlated all of our products to the academic standards of all 50 United States, the District of Columbia, the Department of Defense Dependent Schools, and all Canadian provinces. We have also correlated to the Common Core State Standards.

How To Find Standards Correlations

To print a customized correlation report of this product for your state, visit our website at **http://www.shelleducation.com** and follow the on-screen directions. If you require assistance in printing correlation reports, please contact Customer Service at 1-877-777-3450.

Purpose and Intent of Standards

Legislation mandates that all states adopt academic standards that identify the skills students will learn in kindergarten through grade twelve. Many states also have standards for Pre-K. This same legislation sets requirements to ensure the standards are detailed and comprehensive.

Standards are designed to focus instruction and guide adoption of curricula. Standards are statements that describe the criteria necessary for students to meet specific academic goals. They define the knowledge, skills, and content students should acquire at each level. Standards are also used to develop standardized tests to evaluate students' academic progress. Teachers are required to demonstrate how their lessons meet state standards. State standards are used in the development of all of our products, so educators can be assured they meet the academic requirements of each state.

McREL Compendium

We use the Mid-continent Research for Education and Learning (McREL) Compendium to create standards correlations. Each year, McREL analyzes state standards and revises the compendium. By following this procedure, McREL is able to produce a general compilation of national standards. Each lesson in this product is based on one or more McREL standards, which are listed on the Teacher Resource CD (mcrel.pdf).

TESOL Standards

The lessons in this book promote English language development for English language learners. The standards listed on the Teacher Resource CD (tesol.pdf) support the language objectives presented throughout the lessons.

Common Core State Standards

The lessons in this book are aligned to the Common Core State Standards (CCSS). The standards listed on pages 27–31 (ccss.pdf) support the objectives presented throughout the lessons.

NCTM Standards

The lessons in this book are aligned to the National Council of Teachers of Mathematics (NCTM) standards. The standards listed on the Teacher Resource CD (nctm.pdf) support the objectives presented throughout the lessons.

Correlations to Standards (cont.)

Common Core State Standards Correlation

Common Core Standard	Lesson
3.OA.1 Interpret products of whole numbers, e.g., interpret 5 × 7 as the total number of objects in 5 groups of 7 objects each. For example, describe a context in which a total number of objects can be expressed as 5 × 7.	Snack Time, page 32; Floor Tiles, page 34; Boxes of Cupcakes, page 38; Pattern Questions, page 40; First Names, page 42; Pose a Problem, page 44; Pattern Hunt, page 48; Boxes and Boxes, page 50; Figure It, page 52; At the Fair, page 54; Yard Sale, page 56; What's Going On?, page 58; Buildopoly, page 62; Animal Facts, page 68; Town Races, page 74; How Much Money?, page 86; Classroom Data, page 100; Which One?, page 108; Hobbies, page 110; Table Shapes, page 122
3.OA.2 Interpret whole-number quotients of whole numbers, e.g., interpret 56 ÷ 8 as the number of objects in each share when 56 objects are partitioned equally into 8 shares, or as a number of shares when 56 objects are partitioned into equal shares of 8 objects each. For example, describe a context in which a number of shares or a number of groups can be expressed as 56 ÷ 8.	Equal Groups, page 36; Boxes of Cupcakes, page 38; Pattern Questions, page 40; Pose a Problem, page 44; Finish the Steps, page 46; Figure It, page 52; What's Going On?, page 58
3.OA.3 Use multiplication and division within 100 to solve word problems in situations involving equal groups, arrays, and measurement quantities, e.g., by using drawings and equations with a symbol for the unknown number to represent the problem.	Snack Time, page 32; Floor Tiles, page 34; Equal Groups, page 36; Boxes of Cupcakes, page 38; Boxes and Boxes, page 50; Figure It, page 52; At the Fair, page 54; Yard Sale, page 56; Buildopoly, page 62; Town Races, page 74
3.OA.4 Determine the unknown whole number in a multiplication or division equation relating three whole numbers. For example, determine the unknown number that makes the equation true in each of the equations 8 × ? = 48, 5 = _ ÷ 3, 6 × 6 = ?	Floor Tiles, page 34; Equal Groups, page 36; Boxes of Cupcakes, page 38; Pattern Hunt, page 48; Figure It, page 52
3.OA.5 Apply properties of operations as strategies to multiply and divide. Examples: If 6 × 4 = 24 is known, then 4 × 6 = 24 is also known. (Commutative property of multiplication.) 3 × 5 × 2 can be found by 3 × 5 = 15, then 15 × 2 = 30, or by 5 × 2 = 10, then 3 × 10 = 30. (Associative property of multiplication.) Knowing that 8 × 5 = 40 and 8 × 2 = 16, one can find 8 × 7 as 8 × (5 + 2) = (8 × 5) + (8 × 2) = 40 + 16 = 56. (Distributive property.)	Snack Time, page 32; Floor Tiles, page 34; Equal Groups, page 36; Boxes of Cupcakes, page 38; Pattern Hunt, page 48

Operations and Algebraic Thinking

Correlations to Standards (cont.)

Common Core State Standards Correlation (cont.)

Common Core Standard	Lesson
3.OA.6 Understand division as an unknown-factor problem. For example, find 32 ÷ 8 by finding the number that makes 32 when multiplied by 8.	Equal Groups, page 36; Boxes of Cupcakes, page 38; Pattern Questions, page 40; Figure It, page 52
3.OA.7 Fluently multiply and divide within 100, using strategies such as the relationship between multiplication and division (e.g., knowing that 8 × 5 = 40, one knows 40 ÷ 5 = 8) or properties of operations. By the end of Grade 3, know from memory all products of two one-digit numbers.	Snack Time, page 32; Floor Tiles, page 34; Equal Groups, page 36; Boxes of Cupcakes, page 38; First Names, page 42; Pose a Problem, page 44; Boxes and Boxes, page 50; Figure It, page 52; At the Fair, page 54; Yard Sale, page 56; What's Going On?, page 58; Buildopoly, page 62
3.OA.8 Solve two-step word problems using the four operations. Represent these problems using equations with a letter standing for the unknown quantity. Assess the reasonableness of answers using mental computation and estimation strategies including rounding.	Floor Tiles, page 34; Equal Groups, page 36; Boxes of Cupcakes, page 38; Pose a Problem, page 44; Boxes and Boxes, page 50; At the Fair, page 54; Yard Sale, page 56; What's Going On?, page 58; Buildopoly, page 62
3.OA.9 Identify arithmetic patterns (including patterns in the addition table or multiplication table), and explain them using properties of operations. *For example, observe that 4 times a number is always even, and explain why 4 times a number can be decomposed into two equal addends.*	Pattern Questions, page 40; Pattern Hunt, page 48
3.NBT.1 Use place value understanding to round whole numbers to the nearest 10 or 100.	Number Models, page 60; Animal Facts, page 68; Family Trips, page 70
3.NBT.2 Fluently add and subtract within 1000 using strategies and algorithms based on place value, properties of operations, and/or the relationship between addition and subtraction.	Number Models, page 60; Buildopoly, page 62; Toy Store, page 64; Some Sums, page 66; Animal Facts, page 68; Family Trips, page 70; Make It True, page 72; Town Races, page 74; How Much Money?, page 86; Balance It, page 94; What Does It Hold?, page 96
3.NBT.3 Multiply one-digit whole numbers by multiples of 10 in the range 10–90 (e.g., 9 × 80, 5 × 60) using strategies based on place value and properties of operations.	Buildopoly, page 62; Town Races, page 74; How Much Money?, page 86

Operations and Algebraic Thinking (cont.)

Number and Operations in Base Ten

Correlations to Standards (cont.)

Common Core State Standards Correlation (cont.)

Common Core Standard	Lesson
3.NF.1 Understand a fraction 1/*b* as the quantity formed by 1 part when *a* whole is partitioned into *b* equal parts; understand a fraction *a/b* as the quantity formed by a parts of size 1/*b*.	Make It Match, page 76; On the Number Line, page 78; On the Trail, page 80; Standing in Line, page 82; Who Is Where?, page 84; How Much Money?, page 86; Measure It, page 102
3.NF.2 Understand a fraction as a number on the number line; represent fractions on a number line diagram.	Make It Match, page 76; On the Number Line, page 78; Who Is Where?, page 84; Measure It, page 102
3.NF.2a Represent a fraction 1/*b* on a number line diagram by defining the interval from 0 to 1 as the whole and partitioning it into *b* equal parts. Recognize that each part has size 1/*b* and that the endpoint of the part based at 0 locates the number 1/*b* on the number line.	Make It Match, page 76; On the Number Line, page 78; Who Is Where?, page 84
3.NF.2b Represent a fraction *a/b* on a number line diagram by marking off a lengths 1/*b* from 0. Recognize that the resulting interval has size *a/b* and that its endpoint locates the number *a/b* on the number line.	Make It Match, page 76; On the Number Line, page 78; Who Is Where?, page 84; Measure It, page 102
3.NF.3 Explain equivalence of fractions in special cases, and compare fractions by reasoning about their size.	How Much Money?, page 86; What Is the Number?, page 88; Parts of Shapes, page 130
3.NF.3a Understand two fractions as equivalent (equal) if they are the same size, or the same point on a number line.	Who Is Where?, page 84; What Is the Number?, page 88; Measure It, page 102
3.NF.3b Recognize and generate simple equivalent fractions, e.g., 1/2 = 2/4, 4/6 = 2/3). Explain why the fractions are equivalent, e.g., by using a visual fraction model.	Who Is Where?, page 84; What Is the Number?, page 88
3.NF.3c Express whole numbers as fractions, and recognize fractions that are equivalent to whole numbers. *Examples: Express 3 in the form 3 = 3/1; recognize that 6/1 = 6; locate 4/4 and 1 at the same point of a number line diagram.*	What Is the Number?, page 88
3.NF.3d Compare two fractions with the same numerator or the same denominator by reasoning about their size. Recognize that comparisons are valid only when the two fractions refer to the same whole. Record the results of comparisons with the symbols >, =, or <, and justify the conclusions, e.g., by using a visual fraction model.	On the Trail, page 80; Standing in Line, page 82; Who Is Where?, page 84; What Is the Number?, page 88; Measure It, page 102

(Left vertical label: Number and Operations—Fractions)

Correlations to Standards (cont.)

Common Core State Standards Correlation (cont.)

Common Core Standard	Lesson
3.MD.1 Tell and write time to the nearest minute and measure time intervals in minutes. Solve word problems involving addition and subtraction of time intervals in minutes, e.g., by representing the problem on a number line diagram.	Saturday Mornings, page 90; Pose the Question, page 92; Moving Along, page 98
3.MD.2 Measure and estimate liquid volumes and masses of objects using standard units of grams (g), kilograms (kg), and liters (l).[1] Add, subtract, multiply, or divide to solve one-step word problems involving masses or volumes that are given in the same units, e.g., by using drawings (such as a beaker with a measurement scale) to represent the problem.	Balance It, page 94; What Does It Hold?, page 96; Moving Along, page 98
3.MD.3 Draw a scaled picture graph and a scaled bar graph to represent a data set with several categories. Solve one- and two-step "how many more" and "how many less" problems using information presented in scaled bar graphs. *For example, draw a bar graph in which each square in the bar graph might represent 5 pets.*	Classroom Data, page 100; Keeping Track, page 104
3.MD.4 Generate measurement data by measuring lengths using rulers marked with halves and fourths of an inch. Show the data by making a line plot, where the horizontal scale is marked off in appropriate units— whole numbers, halves, or quarters.	Measure It, page 102
3.MD.5a Recognize area as an attribute of plane figures and understand concepts of area measurement. A square with side length 1 unit, called "a unit square," is said to have "one square unit" of area, and can be used to measure area.	Make It Yourself, page 106; Which One?, page 108; Hobbies, page 110; All Around the Garden, page 112; Many Measures, page 114
3.MD.6 Measure areas by counting unit squares (square cm, square m, square in, square ft, and improvised units)	Make It Yourself, page 106; Hobbies, page 110
3.MD.7b Relate area to the operations of multiplication and addition. Multiply side lengths to find areas of rectangles with whole-number side lengths in the context of solving real world and mathematical problems, and represent whole-number products as rectangular areas in mathematical reasoning.	Which One?, page 108; Hobbies, page 110
3.MD.8 Solve real-world and mathematical problems involving perimeters of polygons, including finding the perimeter given the side lengths, finding an unknown side length, and exhibiting rectangles with the same perimeter and different areas or with the same area and different perimeters.	Make It Yourself, page 106; All Around the Garden, page 112; Many Measures, page 114

Measurement and Data

Correlations to Standards *(cont.)*

Common Core State Standards Correlation *(cont.)*

	Common Core Standard	Lesson
Geometry	**3.G.1** Understand that shapes in different categories (e.g., rhombuses, rectangles, and others) may share attributes (e.g., having four sides), and that the shared attributes can define a larger category (e.g., quadrilaterals). Recognize rhombuses, rectangles, and squares as examples of quadrilaterals, and draw examples of quadrilaterals that do not belong to any of these subcategories.	Tangram Shapes, page 116; Tell Me More, page 118; Name the Shape, page 120; Table Shapes, page 122; Shape Sentences, page 124; They Belong Together, page 126; Draw Me, page 128; Parts of Shapes, page 130
	3.G.2 Partition shapes into parts with equal areas. Express the area of each part as a unit fraction of the whole. For example, partition a shape into 4 parts with equal area, and describe the area of each part as 1/4 of the area of the shape.	Parts of Shapes, page 130

Snack Time

Standards
- Multiplies whole numbers
- Understands strategies for the multiplication of whole numbers

Overview
These problems emphasize the "equal groups" meaning of multiplication. Students have the opportunity to deepen their understanding of equal groups as they consider different ways to find the answers.

Problem-Solving Strategies
- Count, compute, or write an equation
- Organize information in a picture, list, table, graph, or diagram

Materials
- *Snack Time* (page 33; snacktime.pdf)
- *Student Response Form* (page 132; studentresponse.pdf) *(optional)*

Activate
1. Display the following problem for students: *There are 3 plates of oranges. There are 6 oranges on each plate. You eat one of the oranges. How many oranges are left?* (17) Have students work with a partner to find the answer.

2. Have partners tell the larger group how they solved the problem. Encourage a variety of responses. If equations are not included in student explanations, ask *How could we write an equation to show what you did?* Record the equations so that students can refer to these ideas as they work. Possible equations include: $6 + 6 + 6 - 1$, $3 \times 6 - 1$, $6 + 6 + 5$, and $2 \times 6 + 5$, but students may only identify one or two at this time.

Solve
1. Distribute copies of *Snack Time* to students. Have students work alone, in pairs, or in small groups.

2. As appropriate, encourage students to find more than one way to find the answer.

3. Before debriefing have students place their work on their desks and take a gallery walk to see what strategies others used and how they recorded their thinking.

Debrief
1. What answer did you find? How did you find it? What equation did you (could we) write for this way of thinking?

2. Who can tell us a different way to think about this problem?

3. How are addition and multiplication related? Can you explain it using our equations?

Differentiate ◯
Some students may exclusively rely on addition to find answers. Provide them with more opportunities to connect addition and multiplication equations and continue to expose them to different ways of thinking.

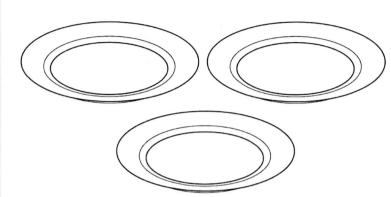

There are 6 carrot sticks on each plate.

How many carrot sticks are there in all?

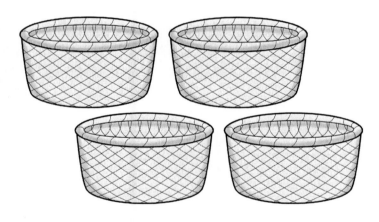

There are 5 apples in each basket.

Jamie takes an apple from each basket to give to her friends.

In all, how many apples are left?

There are 9 almonds in each bowl.

Connor takes 2 almonds from each bowl.

In all, how many almonds are left?

Floor Tiles

Standards

- Multiplies whole numbers
- Understands strategies for the multiplication of whole numbers

Overview

These one- and two-step problems about floor tiles emphasize the area model of multiplication.

Problem-Solving Strategies

- Count, compute, or write an equation
- Organize information in a picture, list, table, graph, or diagram

Materials

- *Floor Tiles* (page 35; floortiles.pdf)
- graph paper
- 1-inch tiles (*optional*)
- *Student Response Form* (page 132; studentresponse.pdf) (*optional*)

Activate

1. Ask students what they can tell you about rows and columns. Encourage a variety of responses. If students keep a math journal or notebook, have them make a picture to help them remember that rows are horizontal and columns are vertical.

2. Distribute a sheet of graph paper to each student. Tell students to imagine that they are tiling a floor. They are going to make five rows of tiles and put four tiles in each row. Then have students make a picture of this floor. Once students have indicated a five-by-four rectangle on the graph paper, ask them how many tiles they used in all and what multiplication equation they could write to represent the number of tiles in this floor. *(5 × 4 = 20)*

Solve

1. Distribute copies of *Floor Tiles* to students. Have students work alone, in pairs, or in small groups. Make graph paper available for students to use.

2. Ask clarifying and refocusing questions as students work, such as *What do you need to find first? How will you use the number on the box?*

Debrief

1. How did you find the number of tiles used?

2. What multiplication equation could we write for the number of tiles used?

3. How did you find the number of tiles left in the box? Did anyone find it a different way?

Differentiate ○

Some students may prefer to use 1-inch tiles to represent the floor concretely. You may wish to observe students as they make such models of the floors. Do they count the tiles in each row as they make it, or build to match the length of the first row? Do they skip count, count by ones, or use multiplication strategies to find the total?

Jackson bought this box of tiles.

He made a tile floor in his hallway.

He put the tiles in 3 rows.

He used 5 tiles in each row.

How many tiles did Jackson use in the hallway?

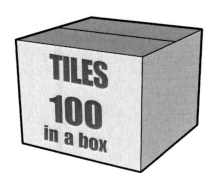

Sophia bought this box of tiles.

She made a tile floor in her kitchen.

She put the tiles in 5 rows.

She used 9 tiles in each row.

How many tiles does Sophia have left?

Peyton bought this box of tiles.

He made a tile floor in his workshop.

He put the tiles in 7 rows.

He used 8 tiles in each row.

Peyton also used 64 of the tiles for a porch floor.

How many tiles does Peyton have left?

Equal Groups

Standards
- Divides whole numbers
- Understands strategies for the multiplication of whole numbers and their division

Overview
Students solve problems about equal groups and division. They find the number in each group or the number of groups. They write equations to represent the problem situations.

Problem-Solving Strategies
- Count, compute, or write an equation
- Organize information in a picture, list, table, graph, or diagram

Materials
- *Equal Groups* (page 37; equalgroups.pdf)
- *Crayon Box Picture* (crayonbox.pdf)
- *Student Response Form* (page 132; studentresponse.pdf) *(optional)*

Activate
1. Show students the *Crayon Box Picture*. Tell them to imagine that all of their crayons are in boxes like this one, and that they have 32 crayons in all. Ask them how many of these boxes of crayons they have. Have students work with a partner to find the answer.

2. Ask partners to tell the larger group how they solved the problem. Encourage a variety of responses. Ask students how they could write an equation to show what they did. Record the equations so that students can refer to these ideas as they work. Possible equations include: $8 + 8 + 8 + 8 = 32$; $8 \times 4 = 32$; $32 \div 8 = 4$.

Solve
1. Distribute copies of *Equal Groups* to students. Have students work alone, in pairs, or in small groups.

2. As students work, ask *Is there a different equation you could write? Can you write an equation that includes division? Are you finding the number of groups or the number in each group?*

Debrief
1. What equation did you write?

2. Who can tell us a different way to think about this problem?

3. How are multiplication and division related? Can you explain it using the equations you wrote?

Differentiate ◯ ▢
Just as some students rely on addition to subtract, some students rely on multiplication to divide. Consider using triangle number fact cards that help connect multiplication and division facts. For example, write each of the numbers 4, 7, and 28 in one corner of a triangle, with the 28 at the top. Students can practice identifying any one of the three numbers when it is covered.

Equal Groups

There are 10 *Captain Underpants* books at the library. There are 5 books on each shelf.

Write an equation to represent this situation.

How many shelves of *Captain Underpants* books are there?

Equal Groups

Nellie did 24 jumping jacks. She did the same number of jumping jacks facing each of the four walls of the gym.

Write an equation to represent this situation.

Tomorrow, Nellie wants to do 36 jumping jacks this same way.

How many jumping jacks should Nellie do facing each wall tomorrow?

Equal Groups

Mr. Lopez buys 91 pencils. He keeps 19 of the pencils for his family. He gives the rest of the pencils to his nieces and nephews. He gives each of the 9 children the same number of pencils.

Write an equation to represent this situation.

How many pencils did each niece and nephew get?

Boxes of Cupcakes

Standards

- Understands strategies for the multiplication of whole numbers and their division
- Knows that a variable is a letter or symbol that stands for one or more numbers

Overview

Within the context of buying boxes of cupcakes, these problems emphasize the array model of multiplication and division. Students write an equation to represent the problems and then solve them.

Problem-Solving Strategies

- Count, compute, or write an equation
- Organize information in a picture, list, table, graph, or diagram
- Guess and check or make an estimate

Materials

- *Boxes of Cupcakes* (page 39; boxescupcakes.pdf)
- *Cupcake Picture* (cupcakepicture.pdf)
- *Student Response Form* (page 132; studentresponse.pdf) *(optional)*

Activate

1. Display or distribute copies of *Cupcake Picture* showing two boxes of cupcakes, one with a 2 × 1 array of cupcakes and one with a 3 × 3 array of cupcakes. Ask *If you buy three packages of the small cupcakes, how many cupcakes do you buy? (16) How did you find that answer? Did anyone find it a different way? What multiplication equation could we write to represent this situation? (2 × 3 = x)*

2. Continue to have students respond to questions about the cupcakes and to create equations with variables to represent the different situations. Be sure to include questions that require students to find the number of groups, the number in each group, and the total number. For example, ask *If you only buy large boxes and buy a total of 27 cupcakes, how many large boxes of cupcakes do you buy? (3) What equation could you write to solve this problem? (27 ÷ 9 = c or 9 × b = 27)*

Solve

1. Distribute copies of *Boxes of Cupcakes* to students. Have students work alone, in pairs, or in small groups.

2. Observe students as they work. Are they counting by ones, skip counting, or using multiplication skills? Are they using fact strategies or do they retrieve a known fact?

Debrief

1. What did you get for an answer? How did you find it?

2. Did anyone think about this differently?

3. How do you decide what letter to use to stand for an unknown number?

Differentiate ◯ ☐ △ ☆

To help inform future instruction, have students complete an exit card. Record 5 × 6 for students and ask them to write about how they might find the product.

Kwan bought four of these boxes of cupcakes for the birthday party.

Write an equation to show how many cupcakes Kwan bought. Use a letter to stand for the number you need to find.

How many cupcakes did Kwan buy?

Cassandra bought some of these boxes of cupcakes for the soccer party. She bought 40 cupcakes in all.

Write an equation to show how many cupcakes Cassandra bought. Use a letter to stand for the number you need to find.

How many boxes of cupcakes did Cassandra buy?

Large   Small

Jasmine bought some large boxes of cupcakes to give to friends. Dylan bought some small boxes of cupcakes to give to friends. They both bought the same number of boxes. Together, they bought 84 cupcakes.

Write an equation to show how many boxes of cupcakes they each bought. Use a letter to stand for the number you need to find.

How many boxes of cupcakes did each of them buy?

Boxes of Cupcakes

Boxes of Cupcakes

Boxes of Cupcakes

Pattern Questions

Standards

- Understands the properties of and the relationships among multiplication and division
- Recognizes a wide variety of patterns

Overview

Students examine patterns related to factors and multiples and extend or generalize the patterns to find numbers that are not shown.

Problem-Solving Strategy

Generalize a pattern

Materials

- *Pattern Questions* (page 41; patternquestions.pdf)
- *Student Response Form* (page 132; studentresponse.pdf) *(optional)*

Activate

1. Display the following pattern:

Row 1	1	2	3	4	5
Row 2	6	7	8	9	10
Row 3	11	12	13	14	15
Row 4					

   Point to the first column in row 4 and ask students what number will go in that spot, and why they think so. When students agree the answer is 16, record the number in that position. Repeat for the rest of the row.

2. Have students talk to a partner about what number they think will be the last number in the fifth row. *(25)* When students are ready, have them share their thinking.

3. Repeat step 2, asking about the last number in row 6. *(30)*

Solve

1. Distribute copies of *Pattern Questions* to students. Have students work in pairs or in small groups so they can answer one another's new questions about the pattern.

2. Encourage students to use patterns rather than write out the next rows. Ask them how many numbers are in each row, and how that information could be helpful.

Debrief

1. What multiplication and division patterns did you find?

2. What are different equations that could be written for the pattern?

3. What question did you create?

Differentiate ◯ ☆

Some students may find it challenging to create a new question. It might help them to first make a list of what they notice about the rows. Have them talk about what stays the same and what changes.

Pattern Questions

Row 1	1	2	3
Row 2	4	5	6
Row 3	7	8	9
Row 4	10	11	12
Row 5			
Row 6			

This pattern of numbers continues.

What will be the last number in row 6?

Ask another question about this pattern.

Pattern Questions

Row 1	1	2	3	4
Row 2	5	6	7	8
Row 3	9	10	11	12
Row 4	13	14	15	16

This pattern of numbers continues.

What will be the last number in row 7?

Write an equation to find the last number in row 10.

Ask another question about this pattern.

Pattern Questions

Row 1	1	2	3	4	5	6
Row 2	7	8	9	10	11	12
Row 3	13	14	15	16	17	18
Row 4	19	20	21	22	23	24

This pattern of numbers continues.

What will be the last number in row 8?

Write an equation to find the row number when the last number in the row is 54.

What will be the second number in row 10?

Ask another question about this pattern.

First Names

Standards

- Understands the basic difference between odd and even numbers
- Understands the properties of and the relationships among multiplication and division
- Knows the language of basic operations

Overview

Students use clues to determine the number of letters in a person's first name. The clues emphasize the language of multiplication and missing factors.

Problem-Solving Strategies

- Count, compute, or write an equation
- Use logical reasoning

Materials

- *First Names* (page 43; firstnames.pdf)
- multiplication charts (*optional*)
- *Student Response Form* (page 132; studentresponse.pdf) (*optional*)

Activate

1. Ask students what the word *product* means in mathematics. Have several students respond to support different ways of saying what the term means.

2. Tell students that you are thinking of a number and are going to give them a clue to find it. Say *When I multiply my number by 5, I get a product of 30. What is my number?* (6) Provide another example, such as *When I multiply my number by 7, I get a product of 21. What is my number?* (3)

3. Provide a clue that is not sufficient for finding a solution, such as *When I multiply my number by 2, I get an answer that is between 11 and 15.* When 6 and 7 are identified as the two possible numbers, give the clue *My number is odd.* (7)

Solve

1. Distribute copies of *First Names* to students. Have students work alone, in pairs, or in small groups.

2. Ask questions such as *What numbers do you say when you count by fives? Which of these numbers are between 14 and 24?* (Ask questions using numbers appropriate for on- and above-level problems.)

Debrief

1. How did the clues help you to find the number?

2. If I am multiplying my number by five and my product is even, what do you know about my number?

Differentiate ◯ ▢ △

To focus on the logical reasoning needed to solve these problems, you may wish to provide some students with a copy of a multiplication chart. Students could use the chart as a reference once they have interpreted the meaning of the clues. Other students may enjoy the challenge of creating clues for the number of letters in their names.

First Names ○

How many letters are in my first name?

Clues:

- When you multiply the number of letters by 5, the answer is between 14 and 24.

- The number of letters is even.

How many letters are there?

First Names ▢

How many letters are in my first name?

Clues:

- When you multiply the number of letters by 7, the product is between 36 and 55.

- The number of letters is odd.

How many letters are there?

First Names △

How many letters are in my first name?

Clues:

- The product of the number of letters and 9 is between 46 and 80.

- The number of letters is even.

- The product of the number of letters and 4 is between 27 and 35.

How many letters are there?

Pose a Problem

Standards

- Multiplies and divides whole numbers
- Understands that mathematical ideas and concepts can be represented concretely, graphically, and symbolically

Overview

Students are asked to make connections among multiple representations involving multiplication and division. Shown a diagram, students are asked to write an equation and a story problem to match the diagram, and then to solve the problem they create.

Problem-Solving Strategies

- Find information in a picture, list, table, graph, or diagram
- Count, compute, or write an equation

Materials

- *Pose a Problem* (page 45; poseproblem.pdf)
- *Student Response Form* (page 132; studentresponse.pdf) *(optional)*

Activate

1. Have students make a picture or diagram of the following story problem: *There are 4 bags of pears. There is the same number of pears in each bag. There are 20 pears in all.* Encourage several responses, inviting students to share their representations. Ask *What are some ways our diagrams are alike? How are they different?*

2. If no one suggests a bar model, add the following diagram to the others. Ask students how this diagram represents the situation.

?	?	?	?
20			

3. Ask *What if we knew there were five pears in each bag, but you did not know the number of bags? How would this diagram change?*

Solve

1. Distribute copies of *Pose a Problem* to students. Have students work alone, in pairs, or in small groups.

2. As students work, note those who create equations or problems in the same number fact family, but not ones that directly match the diagram.

Debrief

1. What equations did you write?

2. What story problems did you create?

3. What is similar about these problems?

Differentiate ⬤ ☆

You may want to meet with a small group of students to brainstorm some ideas before they create their own stories. To help students create problems, have them first brainstorm a list of everyday objects that come in equal groups.

Pose a Problem

Youko made this diagram while solving a problem.

Write an equation to fit Youko's diagram.

Write a story problem to match the equation.

Give the answer to your problem.

Pose a Problem

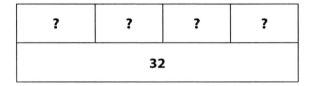

Marcella made this diagram while solving a problem.

Write an equation to fit Marcella's diagram.

Write a story problem to match the equation.

Give the answer to your problem.

Pose a Problem

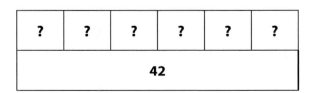

Roberto made this diagram while solving a problem.

Write an equation to fit Roberto's diagram.

Write a story problem to match the equation.

Give the answer to your problem.

Finish the Steps

Standards

- Multiplies and divides whole numbers
- Understands the properties of and the relationships among addition, subtraction, multiplication, and division

Overview

Students are shown a starting number and an ending number. Division and addition must be used to get from the first to last numbers. Students must identify the missing divisor and addend.

Problem-Solving Strategies

- Count, compute, or write an equation
- Guess and check or make an estimate

Materials

- *Finish the Steps* (page 47; finishsteps.pdf)
- *Student Response Form* (page 132; studentresponse.pdf) *(optional)*

Activate

1. Display the following diagram:

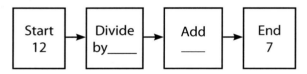

   Have students talk with a neighbor about numbers they could choose to finish the steps of the problem. *(2, 1; 3, 3; 4, 4; 6, 5; 12, 6)*

2. Have students report their answers. Record the numbers as they are named. Ask them what they notice about all of the numbers they used to finish the division step, and whether they think there are any other numbers that could be used.

Solve

1. Distribute copies of *Finish the Steps* to students. Have students work alone, in pairs, or in small groups.

2. As students work, ask clarifying or refocusing questions such as *What pairs of numbers have the start number as a product? Can you show me how you got from this step to this step? Will this step* (pointing to one of the computations) *result in a number that is less or more?*

Debrief

1. What did you find for the missing numbers? How did you find them?

2. Are there other possible answers?

3. Why do these problems have more than one answer? Have we found them all?

4. How can we check our answers?

Differentiate ⬤

Encourage students who are unsure of where to begin to think about which numbers they can multiply to get the starting number as the product.

What are two different ways to finish the steps?

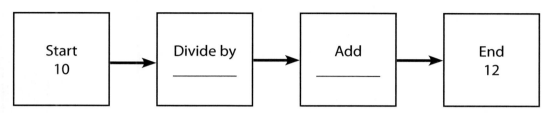

What are three different ways to finish the steps?

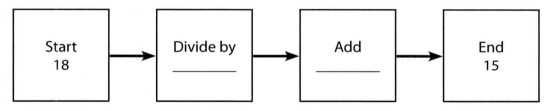

What are all the different ways to finish the steps?

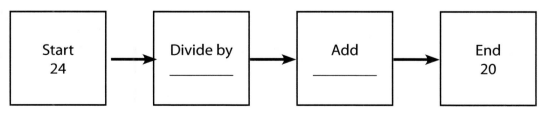

Pattern Hunt

Standards

- Understands the properties of and the relationships among addition, subtraction, multiplication, and division
- Recognizes a wide variety of patterns

Overview

Students examine patterns among a series of equations involving multiplication and addition. They then use the patterns to complete the next equation in the pattern.

Problem-Solving Strategies

- Generalize a pattern
- Use logical reasoning

Materials

- *Pattern Hunt* (page 49; patternhunt.pdf)
- *Student Response Form* (page 132; studentresponse.pdf) *(optional)*

Activate

1. Display the following for students: $0 \times 4 = 0$, $1 \times 4 = 4$, $2 \times 4 = 8$, $3 \times 4 = 12$, $4 \times 4 = 16$, $5 \times 4 = 20$. Ask students how they can use patterns to write the next line, and what this tells them about multiplication and addition. Repeat for other multiplication facts, as necessary.

2. Write the first five multiples of six on the board. *(6, 12, 18, 24, 30)* Tell students that these are products in a multiplication table. Have them talk with a partner about which table it is and why they think so. Have pairs share their thinking with the larger group.

Solve

1. Distribute copies of *Pattern Hunt* to students. Have students work in pairs or in small groups so they can discuss the patterns that they find.

2. Observe students as they work together. Does each member of the pair or group contribute? How do they justify their thinking to one another? Are they interested in finding several patterns, or satisfied with just one or two?

Debrief

1. What patterns did you find?

2. How did you use this information to fill in the next line?

3. What is the relationship between the numbers on the left and the right sides of the equal sign?

Differentiate ◯ ▢ △ ☆

Encourage all students to participate by asking questions that most students can answer, such as *What number is always added in the pattern? With what number does each line in the pattern begin?* To provide more of a challenge for some students, ask them to consider whether the products in each pattern are even or odd, and to explain why that is so.

Pattern Hunt ◯

What patterns do you see?

Use the patterns to decide what comes next.

$2 \times 2 + 1 = 5$

$2 \times 3 + 1 = 7$

$2 \times 4 + 1 = 9$

$2 \times \underline{\hspace{1cm}} + 1 = \underline{\hspace{1cm}}$

Pattern Hunt ▢

What patterns do you see?

Use the patterns to decide what comes next.

$3 \times 3 + 3 = 12 = 4 \times 3$

$3 \times 4 + 3 = 15 = 5 \times 3$

$3 \times 5 + 3 = 18 = 6 \times 3$

$3 \times \underline{\hspace{1cm}} + 3 = \underline{\hspace{1cm}} = \underline{\hspace{1cm}} \times 3$

Pattern Hunt △

What patterns do you see?

Use the patterns to decide what comes next.

$7 \times 1 + 7 = 14 = 2 \times 7$

$7 \times 3 + 7 = 28 = 4 \times 7$

$7 \times 5 + 7 = 42 = 6 \times 7$

$7 \times \underline{\hspace{1cm}} + 7 = \underline{\hspace{1cm}} = \underline{\hspace{1cm}} \times \underline{\hspace{1cm}}$

Boxes and Boxes

Standards

- Multiplies whole numbers
- Understands the properties of and the relationship among addition and multiplication

Overview

In these multistep problems students are told about equal groups of beads, cards, or books in boxes. They then determine the total number of items in the boxes.

Problem-Solving Strategies

- Count, compute, or write an equation
- Organize information in a picture, list, table, graph, or diagram
- Use logical reasoning

Materials

- *Boxes and Boxes* (page 51; boxesboxes.pdf)
- 3 small boxes
- color tiles or cubes (15 red; 12 blue)
- *Student Response Form* (page 132; studentresponse.pdf) *(optional)*

Activate

1. Cluster students around three small boxes. Have them watch as you place five red tiles (or cubes) in each box. Then place four blue tiles (or cubes) in each box. Have each student turn to a neighbor and talk about the total number of tiles (or cubes) in these boxes. Encourage students to find more than one way to decide. Have the pairs report their thinking.

2. Introduce or review the distributive property: $a \times (b + c) = (a \times b) + (a \times c)$. Ask students to explain how the distributive property relates to their thinking.

Solve

1. Distribute copies of *Boxes and Boxes* to students. Have students work alone, in pairs, or in small groups.

2. As students work, note their solution strategies. Do they draw, add, skip count, or multiply to find the totals?

Debrief

1. What total did you find? How did you find it?

2. What is a different way to find the total?

3. What do these different ways tell us about addition and multiplication?

Differentiate ◯ ▢ △ ☆

Drawings allow access to this problem for students who might otherwise not recognize the multiplicative relationships. Encourage these students to identify equal groups and to look at their drawings as they listen to the more abstract thinking of others. If you wish to assign an exit card consider the following problem: *There are 4 boxes of pens. There are 3 black, 2 blue, and 1 red pen in each box. What is the total number of pens in these boxes?*

Boxes and Boxes

There are 3 boxes of beads.

There are 4 silver and 5 gold beads in each box.

What is the total number of beads in these boxes?

Boxes and Boxes

There are 6 boxes of playing cards.

There are 3 queens, 3 kings, and 2 jacks in each box.

What is the total number of cards in these boxes?

Boxes and Boxes

There are 7 large boxes of books. There are 4 mysteries, 2 joke books, and 3 fiction books in each large box.

There are 8 small boxes of books. There are 2 art books and 2 nature books in each small box.

What is the total number of books in these boxes?

Figure It

Standards

- Understands the properties of and the relationships among addition, subtraction, multiplication, and division
- Knows that a variable is a letter or symbol that stands for one or more numbers

Overview

This problem set presents equations with shapes as place holders for numbers. Students must identify numbers that make the equations true.

Problem-Solving Strategies

- Count, compute, or write an equation
- Use logical reasoning
- Guess and check or make an estimate
- Work backward

Materials

- *Figure It* (page 53; figureit.pdf)
- *Student Response Form* (page 132; studentresponse.pdf) (*optional*)

Activate

1. Display the equation $\square \times \square = 16$ for students and tell them that each square stands for the same number. Ask students how they could figure out the value of the square. Several students may recognize that $4 \times 4 = 16$ and say they just know the answer. Probe deeper by asking what they could suggest to a friend who did not recognize this fact.

2. Display the equation $\square + \bigcirc = 31$ for students. Ask *If the square is equal to four, what is the value of the circle?* (27) Have students talk with partners about this and then share as a group.

Solve

1. Distribute copies of *Figure It* to students. Explain to students that they will find the values of the shapes. Remind them that the same shape represents the same number.

2. Have students work in pairs or in small groups so they can talk about their thinking. As they work, pay particular attention to how students communicate about the relationships among addition, subtraction, multiplication, and division.

Debrief

1. How did you find the answer?

2. Could there be another number that would work?

3. How can making an incorrect guess help you to make a better guess next time?

Differentiate ⬤

Some students repeatedly make random guesses. Ask questions to help students recognize what they can learn from their incorrect guesses. For example, say *I see you guessed five. Did you get a product that was too small or too large with that guess? Will your next guess be less than five or greater than five?* Other students may be reluctant to guess and thus do not get the opportunity to learn from their guesses. Encourage such students by saying *Let's just pick a number and try it. Maybe we can learn something from our guess.*

Figure It ○

Find the value of each symbol. The same symbols have the same values.

$$\text{☺} \times \text{☺} = 36$$

$$\text{☺} + \triangle + \triangle + \triangle = 21$$

$$\text{☺} = \underline{\quad\quad} \qquad \triangle = \underline{\quad\quad}$$

Figure It □

Find the value of each symbol. The same symbols have the same values.

$$\text{⊠} \div \text{⊕} = 8$$

$$\text{⊕} \times \text{⊕} = 49$$

$$\text{⊕} = \underline{\quad\quad} \qquad \text{⊠} = \underline{\quad\quad}$$

Figure It △

Find the value of each symbol. The same symbols have the same values.

$$\bullet + \bullet + \bullet + \bullet + 4 = \star$$

$$\star \div \text{✿} = 4$$

$$\text{✿} \times \text{✿} = 81$$

$$\bullet = \underline{\quad\quad} \qquad \star = \underline{\quad\quad} \qquad \text{✿} = \underline{\quad\quad}$$

At the Fair

Standards
- Multiplies whole numbers
- Understands the properties of and the relationships among addition, subtraction, multiplication, and division

Overview
In this problem set students write missing numbers so that stories make sense.

Problem-Solving Strategies
- Count, compute, or write an equation
- Guess and check or make an estimate
- Use logical reasoning

Materials
- *At the Fair* (page 55; atfair.pdf)
- *Student Response Form* (page 132; studentresponse.pdf) *(optional)*

Activate
1. Display the following problem:

   Use each of the numbers in the box below once in the story. Write the number that fits best on each line.

4	24	6

   There are _____ cups on the Teacup Spinner ride. There are seats for _____ people in each cup. There are _____ seats in all.

2. Have students talk with a partner about where the numbers in the box could go in the story so that it makes sense.

3. Encourage volunteers to share their responses. Make sure students note that the 4 and 6 may be placed on either of the first two lines, while the 24 must be placed on the third line to indicate the total. Ask *Why does the story still make sense when you reverse the order of the 4 and 6?*

Solve
1. Distribute copies of *At the Fair* to students. Have students work alone, in pairs, or in small groups.

2. Ask clarifying or refocusing questions if students appear to struggle with choosing appropriate number placement.

Debrief
1. How did you decide where to place the numbers?

2. What is a different way to write the numbers so the story makes sense? Why do you think there is more than one way?

3. What would you have to think about to make a new set of numbers for the story?

Differentiate ◯ ▢ △ ☆
Some students may detect the relationships among the data by reading the story; others may prefer to focus on the numbers themselves. Either is fine as a starting point.

At the Fair

Use each of the numbers in the box below once in the story. Arrange the numbers so that the story makes sense.

6	30	18	3

Aaron went to the fair on May _____. He bought _____ packs of tickets for rides.

Each pack cost $ _____. Aaron spent $ _____ on these tickets.

Use each of the numbers in the box below once in the story. Arrange the numbers so that the story makes sense.

18	6	39	9	15	4

The Ito family paid $ _____ for each of its _____ members to visit the fair. They

also paid $ _____ for parking their car. They paid a total of $ _____. Rina won

two second prizes for her cucumbers. Each prize was worth $ _____, for a total of

$ _____ in prize money.

Use each of the numbers in the box below once in the story. Arrange the numbers so that the story makes sense.

6	4	7	97	43	9

Mr. Wilson entered the chocolate chip cookie contest for the _____ th year. He

made a total of _____ chocolate chip cookies. He gave _____ chocolate chip

cookies to each of the _____ judges to taste. He saved the other _____ cookies

for his family to eat. His _____ children love his chocolate chip cookies.

Yard Sale

Standards

- Multiplies whole numbers
- Understands the properties of and the relationships among addition, subtraction, multiplication, and division

Overview

Students are shown items for sale along with their prices. They are also told the total number of items a customer bought and the total amount that was spent. Students must determine what items were bought.

Problem-Solving Strategies

- Count, compute, or write an equation
- Guess and check or make an estimate
- Use logical reasoning

Materials

- *Yard Sale* (page 57; yardsale.pdf)
- *Sale Picture* (salepicture.pdf)
- multiplication charts (*optional*)
- *Student Response Form* (page 132; studentresponse.pdf) (*optional*)

Activate

1. Build context for the problem by asking students whether their family or neighbors have ever had a yard sale. Ask students why they think people like to buy things at yard sales.

2. Display *Sale Picture* for students. The picture shows that dolls cost $2 and soccer balls cost $5. Ask *How could you spend $12? (2 soccer balls and 1 doll or 6 dolls) Is there another way? Can you buy only dolls and spend $9? Why or why not? (No; the total number of dolls must be a multiple of 2.)*

Solve

1. Distribute copies of *Yard Sale* to students. Have students work alone, in pairs, or in small groups.

2. Ask questions to probe students' thinking, such as *What number are you trying? Why do you think that number will work? How do you keep track of what you have tried?*

Debrief

1. How did you decide what the customer bought?

2. Did anyone do it a different way?

3. Is there a way to limit the guesses you need to make?

Differentiate ⬤

Students often complete several mental computations before finding the correct answer. Some students may become frustrated if their fact knowledge does not allow them to recall products easily. You may want to provide those students with a copy of a multiplication chart to support their ability to succeed.

ard Sale ⭕

Block Sets	$4.00 each
Books	$2.00 each
Puzzles	$1.00 each

Hector bought four things.

He spent $13.00.

What did Hector buy?

Yard Sale ⬜

Bats	$3.00 each
Baseballs	$3.00 each
Mitts	$6.00 each

Mr. Johnson bought four things.

He spent $21.00.

What did Mr. Johnson buy?

ard Sale △

Art Kits	$4.00 each
Board Games	$2.00 each
Building Sets	$9.00 each
Video Games	$7.00 each

Chandra wants to spend $34.00.

How can she do this and buy four things?

How can she do this and buy five things?

What's Going On?

Standards

- Multiplies and divides whole numbers
- Understands the properties of and the relationships among addition, subtraction, multiplication, and division

Overview

Students complete a sequence of computational steps and arrive back at the original number, reinforcing the inverse relationship between multiplication and division.

Problem-Solving Strategies

- Count, compute, or write an equation
- Use logical reasoning

Materials

- *What's Going On?* (page 59; whatsgoingon.pdf)
- carrel or large book
- counters
- *Student Response Form* (page 132; studentresponse.pdf) *(optional)*

Activate

1. Ask students what they know about how multiplication is related to division. Record students' responses on the board.

2. Set up a carrel or book to create a screen between yourself and a student volunteer. Gather other students around the volunteer so that they can see what the volunteer sees, even though you cannot. Direct the volunteer to take 1–5 counters and place them behind the screen. Tell the other students to remember the number of counters they see. Allowing time between each direction, say *Line up these counters one below another. Get more counters so that there are three counters in each row. Take all these counters and put them in three equal groups. Now tell me how many groups you have.* Then dramatize thinking seriously until you announce the number of counters the student started with.

3. You may wish to have students discuss with partners how you were able to tell the original number of counters, but do not debrief as a whole group. Many students will have more ideas after solving a similar problem.

Solve

1. Distribute copies of *What's Going On?* to students. Have students work alone, in pairs, or in small groups.

2. Observe whether students immediately recognize that the answers will be the same as the starting number, compute to make the discovery, or appear surprised.

Debrief

1. Why do the numbers end up the same? Can anyone explain that differently?

2. How do these answers relate to the list we made about multiplication and division?

3. What problem could you create that would be like these?

Differentiate ◯ ☆

To help some students better understand the concept of *inverse*, lead a discussion about things we do and undo every day, for example, turning lights on and off or opening and closing doors.

What's Going On?

Start with 3.

Step 1: Multiply by 5.

Step 2: Divide by 5.

What number do you get?

Choose a different start number that is less than 10. Follow the steps in the box to the left.

What number do you get?

Why do you think this happens?

What's Going On?

Start with 6.

Step 1: Add 2.

Step 2: Multiply by 4.

Step 3: Divide by 4.

Step 4: Subtract 2.

What number do you get?

Choose a different start number that is less than 10. Follow the steps in the box to the left.

What number do you get?

Why do you think this happens?

What's Going On?

Start with 8.

Step 1: Multiply by 3.

Step 2: Double it.

Step 3: Divide by six.

What number do you get?

Choose a different start number that is less than 10. Follow the steps in the box to the left.

What number do you get?

Why do you think this happens?

Number Models

Standards

- Understands the basic meaning of place value
- Uses models to identify, order, and compare numbers

Overview

Students are shown a collection of base-ten blocks and asked what numbers they could show that round to a specified ten or hundred.

Problem-Solving Strategies

- Find information in a picture, list, table, graph, or diagram
- Organize information in a picture, list, table, graph, or diagram

Materials

- *Number Models* (page 61; numbermodels.pdf)
- base-ten blocks
- *Number Line 0–100* (numberline100.pdf)
- *Number Line 0–450* (numberline450.pdf)
- *Student Response Form* (page 132; studentresponse.pdf) (*optional*)

Activate

1. Distribute four hundreds, six tens, and nine ones to each group of three to five students. As a group, have students take two of the tens and seven of the ones. Ask them what numbers they can show with these pieces.

2. Once students agree that they can show the numbers 1–7, 10–17, and 20–27 ask them which of those numbers round to 20.

Solve

1. Distribute copies of *Number Models* to students. Keep the base-ten blocks available for student use. Have students work alone, in pairs, or in small groups.

2. To help students self-correct, ask *Do you have enough ones to write that number? Do numbers have to be less than 120 to round to that number?*

Debrief

1. How did you find your answers?

2. Do you think you found all the numbers? Why or why not?

3. How many numbers round to every ten?

Differentiate ⬤ ▢

If some students require more structure for organizing data, you might start a table for them (with headings of hundreds, tens, and ones, or tens and ones), which they can complete. Also, make number-line templates available for students who would benefit from this ordered representation of numbers. Students may debate whether numbers such as 50 and 120 should be included in a count as they do not need to be rounded. Make sure their reasoning is consistent with their responses.

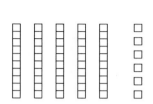

Kate has these tens and ones.

What numbers can she show that round to 50 when rounded to the nearest ten?

Write the numbers.

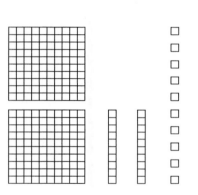

Leo has these hundreds, tens, and ones.

What numbers can he show that round to 220 when rounded to the nearest ten?

Write the numbers.

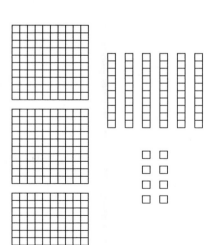

Brianna and Jackson have these hundreds, tens, and ones.

Brianna writes all the numbers she can show that round to 350 when rounded to the nearest ten.

Jackson writes all the numbers he can show that round to 400 when rounded to the nearest hundred.

Who wrote more numbers?

How many more?

Buildopoly

Standards

- Understands the basic meaning of place value
- Multiplies and divides whole numbers
- Solves real-world problems involving number operations, including those that specify units

Overview

Within the context of playing games, students use addition, counting, multiplication, or place value skills to combine items with values of 100, 10, and 1.

Problem-Solving Strategies

- Organize information in a picture, list, table, graph, or diagram
- Use logical reasoning

Materials

- *Buildopoly* (page 63; buildopoly.pdf)
- place-value charts *(optional)*
- *Student Response Form* (page 132; studentresponse.pdf) *(optional)*

Activate

1. Ask students if they have ever played a computer or board game where they built or bought buildings or hotels. Allow a few students to respond. If no one identifies the game Monopoly™, do so yourself. To further activate the context of these problems ask students whether they have played a computer or video game with different levels of challenge, and how those levels differ. Again, allow several students to respond.

2. Have students imagine they are playing a game in which it takes 100 points to build a hotel. Ask *If they built three hotels, how many points would they use?* (300) Ask students to explain how they decided on their answers, and if there is another way to find the answer.

Solve

1. Distribute copies of *Buildopoly* to students. Have students work alone, in pairs, or in small groups.

2. Note whether students record addition or multiplication computations or merely recognize values based on their understanding of place value.

Debrief

1. How do you know your answer is correct?

2. What strategies did you use?

3. Did anyone find the answer another way?

Differentiate ◯ ▢ △ ☆

Some students may benefit from additional practice counting by tens and hundreds. Others may wish to organize the information in a place-value chart. A possible exit card task would be: *Which is greater: 9 tens, 4 ones, and 3 hundreds, or 4 hundreds?*

Buildopoly ◯

Gino and Nicola are playing Buildopoly at level 1. They are each trying to get points so they can build. When they get 10 points they can build a house and when they get 1 point they can build a shed. Gino built 4 houses and 4 sheds. Nicola built 3 houses and 15 sheds.

Who used more points?

How many more points?

Buildopoly ▢

Ursula and Marcel are playing Buildopoly at level 2. They are each trying to get points so they can build. When they get 100 points they can build a mall and when they get 10 points they can build a store. Ursula built 5 malls and 9 stores. Marcel built 6 malls.

Who used more points?

How many more points?

Buildopoly △

Aisha and Hank are playing Buildopoly at level 3. They are each trying to get points so they can build. When they get 100 points they can build a train station, when they get 10 points they can build a bus stop, and when they get 1 point they can build a parking lot. Aisha built 2 train stations, 12 bus stops, and 2 parking lots. Hank built 3 train stations and 17 parking lots.

Who used more points?

How many more points?

Toy Store

Standards

- Performs basic mental computations
- Solves real-world problems involving number operations, including those that specify units

Overview

These problems require students to work backward to find the number of toys that were sold or ordered.

Problem-Solving Strategies

- Work backward
- Organize information in a picture, list, table, graph, or diagram
- Simplify the problem

Materials

- *Toy Store* (page 65; toystore.pdf)
- *Student Response Form* (page 132; studentresponse.pdf) *(optional)*

Activate

1. Ask students what they might find for sale in a toy store. Allow a few students to respond.

2. Display the following problem: *The toy store sold 116 more ships than castles this month. It sold 102 castles. How many ships did it sell? (218 ships)*

3. Have students work in pairs to find the answer. You may wish to challenge students to use only mental computation.

Solve

1. Distribute copies of *Toy Store* to students. Have students work alone, in pairs, or in small groups.

2. Observe students as they work. How do they keep track of their thinking? Do they use mental arithmetic or written algorithms?

Debrief

1. What answer did you find? How did you find it?

2. Why might some people call the numbers in the problem "friendly"?

3. What other problems have you solved like these?

Differentiate ◯ ▢ △

When working backward, some students may need help to organize the information they find. Have them make a list of each item sold. As they find it, they can then write the number sold beside the item name. Students may want to simplify the on- and above-level problems by thinking about the missing question that must be answered first.

Toy Store ◯

Last week Troy's Toy Store sold 145 more jump ropes than kites. It sold 20 kites.

How many jump ropes did the toy store sell last week?

oy Store ▢

Last month Troy's Toy Store sold 223 more baseballs than bats. It sold 130 more bats than mitts. Troy's Toy Store sold 70 mitts.

How many baseballs did the toy store sell last month?

Toy Store △

Today Troy's Toy Store ordered 217 more puzzles than card games. It ordered 245 more card games than craft kits. It ordered 155 craft kits.

How many puzzles did the toy store order today?

Some Sums

Standard

Uses specific strategies to estimate computations and to check the reasonableness of computation skills

Overview

Students are provided with a set of digits they must arrange to find a given sum. Use of estimation skills allows students to limit the amount of trial and error that is necessary.

Problem-Solving Strategies

- Guess and check or make an estimate
- Count, compute, or write an equation

Materials

- *Some Sums* (page 67; somesums.pdf)
- index cards with digits 0, 1, 2, 3, and 4 *(optional)*
- *Student Response Form* (page 132; studentresponse.pdf) *(optional)*

Activate

1. Ask students what it means to say a number has three *digits*. Through discussion help students understand that numbers greater than nine are composed of multiple digits. Numbers in the hundreds are three-digit numbers.

2. Display the equation __ __ + __ __ = 108 and have students record it. Then, display the digits 3, 4, 5, and 6. Tell students that they are to write each of these digits in the blanks so the numbers have a sum of 108. Invite students to work independently or in pairs to complete the task. *(43 + 65)*

3. Have students share their thinking. Ask them what helped them to place the numbers correctly.

4. Repeat the process in step 2, changing the sum to 90. *(34 + 56)*

Solve

1. Distribute copies of *Some Sums* to students. Have students work alone, in pairs, or in small groups.

2. Ask clarifying and refocusing questions as students work. For example, ask *How could estimation help you? What do you know about the numbers in the ones place?*

Debrief

1. How did you place the numbers to find the sum? Did anyone do it differently?

2. How might estimation help to solve these problems?

3. Is it always possible to find the same sum in more than one way when we have different digits? Why or why not?

Differentiate ⬤

Some students might be more successful moving digit cards around to check different arrangements. Make cards with the digits 0–4 available to them. Some students may also find it easier to think about the problem written in a vertical format. Write it for them that way or have them do so.

Some Sums

Use each of the digits 0, 1, 2, 3, and 4 once.

Where should you write the digits so that the addition equation is true?

_____ _____ _____ + _____ _____ = 415

Some Sums

Use each of the digits 1, 3, 5, 7, and 9 once.

Where should you write the digits so that the addition equation is true?

_____ _____ _____ + _____ _____ = 844

Show another way.

_____ _____ _____ + _____ _____ = 844

Some Sums

Use each of the digits 2, 3, 4, 5, 6, and 7 once.

Where should you write the digits so that the addition equation is true?

_____ _____ _____ + _____ _____ _____ = 693

Show three more ways.

_____ _____ _____ + _____ _____ _____ = 693

_____ _____ _____ + _____ _____ _____ = 693

_____ _____ _____ + _____ _____ _____ = 693

Animal Facts

Standards
- Understands the basic meaning of place value
- Multiplies and divides whole numbers

Overview
Students use knowledge of different representations of the same number and addition, subtraction, and multiplication skills to find the answer to a question about animals.

Problem-Solving Strategies
- Find information in a picture, list, table, graph, or diagram
- Use logical reasoning

Materials
- *Animal Facts* (page 69; animalfacts.pdf)
- *Student Response Form* (page 132; studentresponse.pdf) *(optional)*

Activate
1. Ask students what interesting facts they know about animals. Allow several students to respond. Point out any facts that include quantitative data. If no such facts are included, tell students that some interesting facts about animals include numbers. One example is that the elephant is the only animal that has four knees.

2. Display the following problem:

   About how many muscles are in a cat's ear? Use the clues below. Choose the answer from the box. (32)

18	120	32	146

   Clues:

   The number is less than $100 + 40 + 6$.

   It is greater than 3×6.

   It is not equal to 6×20.

   After each clue ask *What does this clue tell us about the number of muscles?*

Solve
1. Distribute copies of *Animal Facts* to students. Tell students that their task is to use the clues to answer the question. Have students work alone, in pairs, or in small groups.

2. Observe students as they work. Do they add the numbers, estimate, or recognize the expanded forms of numbers? What strategies do they use to multiply a multiple of ten?

Debrief
1. What answer did you find? How did you decide it was that number?

2. What is a different solution strategy?

3. How did you keep track of what you learned from each clue?

Differentiate ○ □ △ ☆
Students with a particular interest in these types of problems may wish to create similar problems using their own animal facts. Have pairs of students write a problem together, exchange it with another pair, and solve the new problem.

About how many bones does a lion have?

Use the clues below. Choose the answer from the box.

64	264	275	260

The number is greater than 9×10.

The number has a 6 in the tens place.

The number is less than $200 + 60 + 4$.

A lion has about _____ bones.

About how many muscles does a caterpillar have in its head?

Use the clues below. Choose the answer from the box.

270	1211	248	334

The number is less than $1,000 + 200 + 10 + 1$.

The number has a 2 in the hundreds place.

The number is not equal to 3×90.

A caterpillar has about _____ muscles in its head.

About how many taste buds does a cat have?

Use the clues below. Choose the answer from the box.

2479	271	473	475	400

The number is not equal to $2,000 + 400 + 70 + 9$.

There is a 4 in the hundreds place.

The number is not equal to 5×80.

The number is less than $1,000 - 526$.

A cat has about _____ taste buds.

Animal Facts

Family Trips

Standards

- Uses specific strategies to estimate computations and to check the reasonableness of computation skills
- Solves real-world problems involving number operations, including those that specify units

Overview

Students are shown signs that give the distance (in miles) to possible destinations. Given the total distance of a trip, rounded to the nearest 10 or 100, students identify the specific locations or exact mileage.

Problem-Solving Strategies

- Guess and check or make an estimate
- Count, compute, or write an equation

Materials

- *Family Trips* (page 71; familytrips.pdf)
- *Number Line 0–100* (numberline100.pdf)
- *Number Line 0–450* (numberline450.pdf)
- *Student Response Form* (page 132; studentresponse.pdf) *(optional)*

Activate

1. Display *Number Line 0–100*. Ask students how they can use the number line to help round 47 to the nearest ten. *(50)* Then ask them to identify other numbers that round to 50. *(45, 46, 48, 49, 51, 52, 53, 54)*

2. Ask students what number they get when they round 96 to the nearest hundred. *(100)* Ask students to round other numbers to the nearest hundred, like 51 and 141. *(100)*

3. Have students talk with a partner about this question: *What numbers round to 100 when rounded to the nearest hundred and round to 90 when rounded to the nearest ten?* *(85, 86, 87, 88, 89, 90, 91, 92, 93, 94)* Provide time for them to share their ideas.

Solve

1. Distribute copies of *Family Trips* to students. Have students work alone, in pairs, or in small groups.

2. Note whether students recognize when they can estimate rather than find exact totals.

3. Encourage students with descriptive feedback as they work. For example, say *I see you are thinking about the different numbers that round to 50.*

Debrief

1. Where is the family traveling? How do you know?

2. How might estimation help you solve these problems?

Differentiate ⬤

After adding to find the distance of a particular trip, some students may be uncertain how to round the number of miles. Make copies of *Number Line 0–100* and *Number Line 0–450* for those students to use. The visual model may allow students to recognize the closest 10 or 100.

Your family drives to Milltown and then to a beach. The total miles driven is less than 70 miles.

Is your family going to Cove Beach or Harbor Beach?

Your family drives to Bay Village and then to a mountain. The total miles driven is less than 250 miles.

Is your family going to Blue Mountain, Long Mountain, or Wild Mountain?

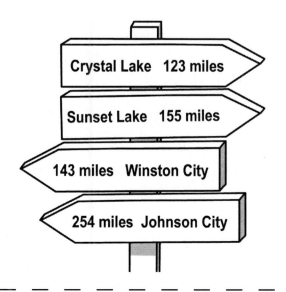

Your family is driving from a city to a lake.

When you round the sum of the miles to the nearest 10 or the nearest 100, the number of miles your family will drive is the same.

How many miles is your family driving?

Family Trips

Make It True

Standard

Uses specific strategies to estimate computation and to check the reasonableness of computation skills

Overview

Students are provided with a set of numbers they must arrange to complete equations. Use of estimation skills allows students to limit the amount of trial and error.

Problem-Solving Strategies

- Guess and check or make an estimate
- Count, compute, or write an equation

Materials

- *Make It True* (page 73; maketrue.pdf)
- *Student Response Form* (page 132; studentresponse.pdf) (*optional*)

Activate

1. Display the following list of numbers: 167, 227, 364, and 75. Have students record the numbers. Direct students to look for two numbers that have a sum of 242. *(167 + 75)* Ask them how they found the numbers. Have several students respond so that a variety of approaches are identified. Though many students may report that they guessed and checked or just found them, probe deeper by asking whether they included 365 in one of their guesses, and why or why not. Have students identify how estimating might help them to find the numbers.

2. Repeat the process in step 1, asking students to find two numbers that have a difference of 137. *(364 – 227)*

Solve

1. Distribute copies of *Make It True* to students. Have students work alone, in pairs, or in small groups.

2. Encourage students as they work with descriptive feedback, such as *I see you are crossing out the numbers you have already used. I hear you rounding the numbers so that you can make better guesses.*

Debrief

1. If a number may only be used once, how can this rule help us to find answers?

2. What equations did you find easiest to complete? What made them easier?

3. Which equations can be completed more than one way? Why?

4. What ideas might we put on a list of ways to help us find these sums and differences?

Differentiate ⬤

Some students may be overwhelmed by all of the possible combinations. In problems where each number can be used only once, encourage students to cross off numbers when they are used. This process will help students focus on the more limited choices for the next example. Consider assigning an exit card task such as the following: *Which two of these numbers have a sum of 410: 243, 257, 167? (243 and 167)*

Use each number in the box only once to complete the equations.

| 348 | 98 | 564 | 34 |

$132 = \underline{\hspace{1.5cm}} + \underline{\hspace{1.5cm}}$

$\underline{\hspace{1.5cm}} - \underline{\hspace{1.5cm}} = 216$

Use each number in the box only once to complete the equations.

| 57 | 195 | 147 | 257 | 746 | 108 |

$342 = \underline{\hspace{1.5cm}} + \underline{\hspace{1.5cm}}$

$\underline{\hspace{1.5cm}} - 489 = \underline{\hspace{1.5cm}}$

$\underline{\hspace{1.5cm}} + 51 = \underline{\hspace{1.5cm}}$

Use the numbers in the box to complete the equations.

| 479 | 295 | 389 | 525 |

$684 = \underline{\hspace{1.5cm}} + \underline{\hspace{1.5cm}}$

$\underline{\hspace{1.5cm}} - 46 = \underline{\hspace{1.5cm}}$

$774 - \underline{\hspace{1.5cm}} = \underline{\hspace{1.5cm}}$

Town Races

Standard
Multiplies and divides whole numbers

Overview
These problems provide practice with multiplying by multiples of ten as students determine the number of children signed up for a particular race. They also use division to find out how many more teams are needed to have a given number of children in the race.

Problem-Solving Strategy
Count, compute, or write an equation

Materials
- *Town Races* (page 75; townraces.pdf)
- *Student Response Form* (page 132; studentresponse.pdf) *(optional)*

Activate

1. To activate prior knowledge ask students what kind of races the school might have at a field day. Encourage students to brainstorm races that involve teams.

2. Ask students how they find 3×20, and why their method works. Demonstrate the associative property of multiplication by showing $3 \times (2 \times 10)$ and $(3 \times 2) \times 10$. Have students discuss how this could help them solve the problem.

3. Display the following problem:

   There are 25 pairs of children signed up for the wheelbarrow race. How many children are signed up for the wheelbarrow race? (50) How many more pairs will have to sign up for this race to have a total of 70 children? (10)

4. Have students work in pairs to solve the problems and then share their thinking with the larger group.

Solve

1. Distribute copies of *Town Races* to students. Have students work alone, in pairs, or in small groups.

2. Ask clarifying and refocusing questions as students work, such as *How did you get this number? What are you trying to do now?*

3. Provide descriptive feedback, such as *I see how you found the number of children here. Your work is labeled clearly.*

Debrief

1. What answer did you find? How did you find it?

2. What answer did you have to find first?

3. What helped you to solve these problems?

Differentiate ◯ ▢ △ ☆
Language can help some students better understand how to multiply with multiples of ten. Understanding that 3×20 can be thought of as 3×2 tens may help some students to recognize the product as 6 tens, or 60. You may want to share this way of thinking with students who are successful as well, to increase their appreciation of our way of recording numbers.

There are 20 pairs of children signed up for the three-legged race.

How many children are signed up for the three-legged race?

How many more pairs would have to sign up for the race to have a total of 50 children?

There are 4 children on each relay team in the egg-on-a-spoon race. There are 30 teams signed up for this race. How many children are signed up for the egg-on-a-spoon race?

How many more teams will have to sign up for the race to have a total of 156 children?

There are 8 children on each relay team in a water balloon race. There are 20 teams signed up for this race.

There are 40 children on each team in the run-backward race. There are 6 teams signed up for this race.

The children who signed up for the water balloon race want to have the same number of racers that signed up for the run-backward race. How many more teams have to sign up for the water balloon race?

Make It Match

Standards

- Understands the concepts related to fractions
- Uses models to identify, order, and compare numbers

Overview

These problems require students to write the fractions indicated on a number line.

Problem-Solving Strategy

Find information in a picture, list, table, graph, or diagram

Materials

- *Make It Match* (page 77; makematch.pdf)
- *Student Response Form* (page 132; studentresponse.pdf) *(optional)*

Activate

1. Ask students what they know about a unit fraction. Collect several ideas and provide time for students to restate the ideas of others. Then, ask *How does knowing what one-third is help us to find two-thirds?*

2. Show the following number line:

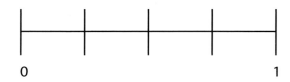

0   1

3. Ask students what the line shows them. Encourage a variety of responses. Then ask students to locate $\frac{1}{4}$ on the line, and to explain how they know it is $\frac{1}{4}$. Repeat this activity with $\frac{3}{4}$.

Solve

1. Distribute copies of *Make It Match* to students. Have students work alone or in pairs.

2. Give descriptive feedback to encourage students. For example, say *I see you are counting the lines carefully.*

Debrief

1. How did you decide on the number to write?

2. What suggestions could you give someone about how to decide which number to write?

Differentiate ⬤ ☆

Some students may describe a fraction such as $\frac{3}{4}$ as "three out of four." Though common, this language will not serve students well when fractions are greater than one. For example, what would $\frac{5}{4}$ mean? Encourage students to think of $\frac{3}{4}$ as three of $\frac{1}{4}$. Consider the following exit card task: *Show $\frac{3}{4}$ on a number line.*

Make It Match

Lei drew an arrow to show how much of a slice of cheese a mouse ate.

What fraction could you write to tell how much of the slice of cheese a mouse ate?

Make It Match

James drew an arrow to show how much of the driveway he shoveled.

What fraction could you write to tell how much of the driveway James shoveled?

Make It Match

Mason made a flag that was yellow and red. More of the flag was red than yellow.

Mason drew arrows to tell how much of the flag was each color.

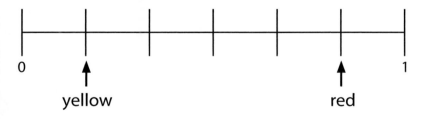

What fraction could you write to tell how much of the flag was red?

What fraction could you write to tell how much of the flag was yellow?

On the Number Line

Standards

- Understands the concepts related to fractions
- Uses models to identify, order, and compare numbers

Overview

These problems ask students to show fractions less than 1 on a number line.

Problem-Solving Strategy

Organize information in a picture, list, table, graph, or diagram

Materials

- *On the Number Line* (page 79; numberline.pdf)
- strips of 4 in. $\times$ $\frac{1}{2}$ in. paper
- *Student Response Form* (page 132; studentresponse.pdf) *(optional)*

Activate

1. Sketch a number line from zero to one with one-half marked incorrectly. For example, show:

   Have students talk with a partner about what they see and then have them share their thinking with the class. Ask students what they need to think about to place the $\frac{1}{2}$ mark where it belongs.

2. Erase the incorrect mark and have a student estimate where it belongs. Then ask for a volunteer to estimate the mark for $\frac{1}{4}$.

Solve

1. Distribute copies of *On the Number Line* to students. Have students work alone, in pairs, or in small groups.

2. To increase the likelihood that students self-correct, ask *Why do you think $\frac{1}{4}$ is placed there? How many one-fourths are there from zero to one?*

Debrief

1. How did you decide where to make the marks?

2. How could you solve the problem in a different way?

3. How might thinking about the order of fractions be helpful?

Differentiate ⬤

Distribute 4-inch strips of paper to students, representing the number line. They can write 0 at the left end and 1 at the right end. To show thirds, for example, students fold the strip into three parts of the same length. Students can then mark each of the thirds on the fold lines and match these to the number line shown in the problem. Alternatively, students can actually fold the number lines shown beneath the problems to find where to place the marks.

Paul ate $\frac{1}{4}$ of a loaf of French bread.

Place a mark on the number line to show how much of the loaf Paul ate. Label the mark with his name.

Marie ate $\frac{1}{3}$ of her long sub sandwich. Jason ate $\frac{2}{3}$ of his sub sandwich.

Place marks on the number line to show how much of their sandwich each person ate. Label each mark with the person's name.

DeShawn, Ruby, and Liam each had a long piece of red licorice. The licorice pieces were all the same length. DeShawn ate $\frac{5}{8}$ of his licorice. Ruby ate $\frac{1}{8}$ of her licorice. Liam ate $\frac{1}{2}$ of his licorice.

Place marks on the number line to show how much of their licorice each person ate. Label each mark with the person's name.

On the Trail

Standards

- Understands the concepts related to fractions
- Uses models to identify, order, and compare numbers

Overview

Students compare fractions that have the same numerator or the same denominator.

Problem-Solving Strategies

- Use logical reasoning
- Organize information in a picture, list, table, graph, or diagram

Materials

- *On the Trail* (page 81; ontrail.pdf)
- about 6 ft. of ribbon or rope
- tape
- *Student Response Form* (page 132; studentresponse.pdf) *(optional)*

Activate

1. Draw a circle and ask students to imagine that it is their favorite cookie, cupcake, or bowl of cereal. Ask them if they were really hungry, whether they would like to eat $\frac{1}{4}$ or $\frac{2}{4}$ of this food. Encourage several students to explain their thinking. Repeat for the numbers $\frac{1}{5}$ and $\frac{3}{5}$.

2. Follow the same procedure as step 1, but this time present fractions with the same numerators, for example, $\frac{1}{5}$ and $\frac{1}{4}$, followed by $\frac{2}{4}$ and $\frac{2}{3}$.

3. Pull the ribbon or rope taut and tape down its ends to the floor. Gather students around the line and ask them to imagine that it is a road. Ask *If you had walked $\frac{1}{2}$ of the distance of this road and your brother had walked $\frac{1}{3}$ of it, who walked more of this road? How do you know?*

Solve

1. Distribute copies of *On the Trail* to students. Have students work alone, in pairs, or in small groups.

2. Observe students as they work. Do they make pictures of the fractions, fold strips of paper, or draw conclusions about fractions with the same numerators or denominators?

Debrief

1. How did you decide who walked more of the trail?

2. What is a different way to think about the problem?

3. How does recognizing that fractions have the same numerator or same denominator help us to compare fractions?

Differentiate

When comparing fractions, thinking about a portion of something they want to eat is easier for some students than thinking about a portion of a length. If students become confused when comparing distances, have them think about sharing something they would like to have.

On the Trail

Jordan and Mika are walking on a nature trail. Jordan has walked $\frac{1}{3}$ of the trail. Mika has walked $\frac{2}{3}$ of the trail. Who has walked more of the trail?

On the Trail

Philip and Rosa are skiing down a ski trail. Philip is $\frac{2}{6}$ of the way down the trail. Rosa is $\frac{2}{8}$ of the way down the trail. Who is farther down the trail?

On the Trail

Sam, Chris, and Kim Su are riding their bikes on a bike trail. Sam has ridden $\frac{3}{4}$ of the trail. Chris has ridden $\frac{3}{8}$ of the trail. Kim Su has ridden $\frac{3}{6}$ of the trail. They want to ride the entire trail. Who has the greatest distance left to ride?

Standing in Line

Standards

- Understands the concepts related to fractions
- Uses models to identify, order, and compare numbers

Overview

These problems are all about people standing in line. Fractions are used to describe part of the line.

Problem-Solving Strategies

- Organize information in a picture, list, table, graph, or diagram
- Use logical reasoning

Materials

- *Standing in Line* (page 83; standingline.pdf)
- *Student Response Form* (page 132; studentresponse.pdf) (*optional*)

Activate

1. Choose six volunteers to stand in line in front of the other students. As you choose students look for a characteristic (such as wearing blue) that some of them share and some of them do not. Ask students what fraction they can use to tell how many of the students are wearing blue. Repeat for other characteristics.

2. Review ordinal numbers by asking who is fourth in line and who is sixth in line.

3. Without a line of people to view, ask questions such as *If you are fifth in line, how many people are in front of you? (4) If there are 7 people in line and you are third, how many people are behind you? (4)*

Solve

1. Distribute copies of *Standing in Line* to students. Have students work alone, in pairs, or in small groups.

2. Ask clarifying or refocusing questions as students work. For example, *Can you show me a picture to help me understand your thinking? How did you indicate yourself in this drawing?*

Debrief

1. How did you find your answer?

2. Who can restate what has been said?

3. How might you help a first-grade student understand how to solve this problem?

Differentiate ◯ ▢ △ ☆

Have students complete an exit card and use their responses to inform your future instructional planning. Pose the problem: *There are six students in line. One-third of the students are boys. How many students are girls?*

Standing in Line ○

What fraction of the people are wearing a hat?

Standing in Line ☐

There are 6 people in line at the bank.

I am third in line.

What fraction of the people in line are in front of me?

Standing in Line △

I am standing in line to buy lunch.

I am third in line.

One-fourth of the people in line are in front of me.

What fraction of the people in line are behind me?

Who Is Where?

Standards
- Understands the concepts related to fractions and decimals
- Uses models to identify, order, and compare numbers

Overview
These problems require students to place fractions in order and then to write a statement about the fractions using >, <, or =.

Problem-Solving Strategies
- Use logical reasoning
- Organize information in a picture, list, table, graph, or diagram

Materials
- *Who Is Where?* (page 85; whowhere.pdf)
- strips of 3 in. × $\frac{1}{2}$ in. paper *(optional)*
- *Student Response Form* (page 132; studentresponse.pdf) *(optional)*

Activate
1. Show students the greater than (>) and less than (<) signs and ask them to write a comparison statement using each sign. Have students share their statements.

2. Display the following list of fractions: $\frac{2}{8}$, $\frac{1}{4}$, $\frac{2}{4}$. This time have students work with partners to write three comparisons of these fractions, one using >, one using <, and one using =. Again, have students share their statements.

3. Ask students how they decided which fractions were equal. Then ask them how they recognized that one fraction was greater than another.

Solve
1. Distribute copies of *Who Is Where?* to students. Have students work alone, in pairs, or in small groups.

2. Ask questions as students work, such as *Why do you think that number goes there? What does the placement of the marks on the street tell you about the size of the numbers?*

Debrief
1. What numbers did you find that are equal? How did you recognize that they were equal?

2. If you know that $\frac{2}{3} > \frac{1}{3}$, what other statement can you write about these two fractions?

Differentiate ⬤
Some students may be able to compare two fractions correctly but unable to correctly place the fractions on the number line. Have these students return to use of paper strips to make the fractions (see *On the Number Line*, page 78). For these problems, use strips that are three inches long.

Who Is Where?

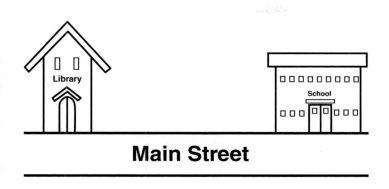

Simon and Nicki are walking down Main Street from their school to the town library. Simon is $\frac{1}{2}$ of the way to the library. Nicki is $\frac{2}{4}$ of the way to the library. Where are they on Main Street? Show them on the map.

Write a statement about these fractions. Use $>$, $<$, or $=$.

Who Is Where?

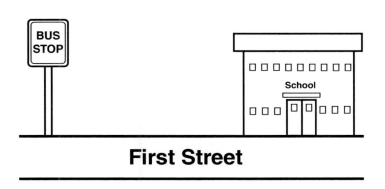

Pam and Robert are walking down First Street from their school to the bus stop. Pam is $\frac{2}{3}$ of the way to the bus stop. Robert is $\frac{4}{6}$ of the way to the bus stop. Where are they on First Street? Show them on the map.

Write a statement about these fractions. Use $>$, $<$, or $=$.

Who Is Where?

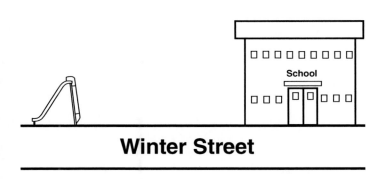

Jackie, Sophie, and Zak are walking down Winter Street from their school to the city park. Jackie is $\frac{3}{4}$ of the way to the park. Sophie is $\frac{3}{8}$ of the way to the park. Zak is $\frac{6}{8}$ of the way to the park. Where are they on Winter Street? Show them on the map.

Write two statements about these fractions. Use $>$, $<$, or $=$.

How Much Money?

Standards

- Understands the concepts related to fractions and decimals
- Uses models to identify, order, and compare numbers

Overview

Students are given information about a group of coins and use that information to determine the total amount of money.

Problem-Solving Strategies

- Organize information in a picture, list, table, graph, or diagram
- Act it out or use manipulatives

Materials

- *How Much Money?* (page 87; howmuchmoney.pdf)
- pictures of U.S. coins
- play or real U.S. coins
- *Student Response Form* (page 132; studentresponse.pdf) *(optional)*

Activate

1. Ask students what they know about the value of U.S. coins. Provide time for several students to respond.

2. Have coins available for students' use. Place two dimes and two pennies in your clasped hands, hiding the coins. Tell students that you have four coins, and that half of them are pennies and half of them are dimes. Ask students what coins you have in your hands.

3. Have students explain how they decided which coins you held. Once several students have explained their thinking, open your hands to reveal your coins. Ask students to find the total value of these coins. *(22¢)*

Solve

1. Distribute copies of *How Much Money?* to students. Have students work alone, in pairs, or in small groups. Have coins available for their use.

2. Provide descriptive feedback to encourage students' thinking. For example, say *I see you are dividing the whole into equal groups. I see you are organizing the information in a table so you can keep track of the number and value of each type of coin.*

Debrief

1. How did you find the number of coins in the fractional part?

2. How did you find the total value of the coins?

Differentiate ☆

Students from other countries may be less familiar with U.S. coins. Provide them with a picture of a penny, nickel, dime, and quarter. Lead them in labeling each coin with its name and value. Point out that the sizes of U.S. coins do not necessarily indicate their relative value (e.g., a dime is smaller than a nickel, but is worth more).

How Much Money? ○

I have 6 coins.

Half of the coins are nickels.

The rest are pennies.

How much money do I have?

How Much Money? ▢

I have 8 coins.

One-fourth of the coins are quarters.

The rest are nickels.

How much money do I have?

How Much Money? △

I have 9 coins.

Two-thirds of the coins are dimes.

The rest are quarters.

How much money will I have left after I buy a hot pretzel for $1.00?

What Is the Number?

Standards

- Understands the concepts related to fractions
- Uses models to identify, order, and compare numbers

Overview

Students use clues to find a number on a sign. Comparison of fractions and common names for the same fraction are emphasized.

Problem-Solving Strategies

- Find information in a picture, list, table, graph, or diagram
- Use logical reasoning

Materials

- *What Is the Number?* (page 89; whatnumber.pdf)
- *Fraction Template* (fractiontemplate.pdf)
- *Student Response Form* (page 132; studentresponse.pdf) *(optional)*

Activate

1. Ask students what they know about the number $\frac{1}{2}$. Have students brainstorm their ideas in small groups.

2. Invite students to share their thinking.

3. Further activate students' prior knowledge by asking them to name fractions that are greater than or less than $\frac{1}{2}$. Ask students what fractions they can name that are equal to $\frac{1}{2}$.

Solve

1. Distribute copies of *What Is the Number?* to students. Point out to students that there are four fractions in each box and that it is their job to use the clues to identify one of them. Have students work alone, in pairs, or in small groups.

2. Ask clarifying and refocusing questions as students work, such as *Why do you think this number is greater? Can you make a picture to prove that it is?*

Debrief

1. How did you use the clues to find the number?

2. What are some other fractions that are greater than one-fourth?

3. How did you note that a number was eliminated by a clue?

4. Can every whole number be written as a fraction? Why?

Differentiate ◯ △

Provide a copy of the fraction template to students who have difficulty comparing fractions. Students can use the visual models to decide if a fraction is more than, less than, or equal to another one. To offer a greater challenge, have students create their own fraction signs and clues.

What is the Number?

$$\frac{1}{2} \quad \frac{1}{6} \quad \frac{2}{3} \quad \frac{3}{3}$$

Find the number in the box.

It is less than $\frac{2}{3}$.

It is greater than $\frac{1}{5}$.

What is the number?

What is the Number?

$$\frac{4}{8} \quad \frac{2}{4} \quad \frac{3}{8} \quad \frac{1}{8}$$

Find the number in the box.

It is greater than $\frac{1}{6}$.

It is not equal to $\frac{1}{2}$.

What is the number?

What is the Number?

$$\frac{2}{1} \quad \frac{4}{4} \quad \frac{2}{8} \quad \frac{5}{8}$$

Find the number in the box.

It is not equal to 1.

It is greater than $\frac{1}{4}$.

It is not equal to 2.

What is the number?

Saturday Mornings

Standard
Understands relationships between measures

Overview
These problems focus on students' abilities to read and write times. The third problem also involves intervals of time.

Problem-Solving Strategies

- Find information in a picture, list, table, graph, or diagram
- Act it out or use manipulatives

Materials
- *Saturday Mornings* (page 91; saturday.pdf)
- analog and digital clock manipulatives
- *Student Response Form* (page 132; studentresponse.pdf) *(optional)*

Activate

1. Engage students in a conversation related to the context of these problems by asking them what time they wake up on Saturday mornings, and whether that time is the same or different from the time they wake up on school days.

2. Invite a few students to share the time they wake up by showing it on an analog or digital clock. Once shown, have students turn to a neighbor and whisper the time. Then, ask someone to write the time. Ask *Should we write A.M. or P.M. after this time?*

3. Show 7:45 on an analog clock and have students read the time. Ask students whether this time is closer to 7:00 or 8:00, and to explain how they know. Invite a student to show a time on a digital clock that is just before 11:00. Have another student name the time that is shown.

Solve

1. Distribute copies of *Saturday Mornings* to students. Have students work alone, in pairs, or in small groups.

2. Ask clarifying questions as students work, such as *How do you find a time that is just before 7:00? How do you know the time this clock shows?*

Debrief

1. What time did you identify?

2. What strategies did you use to solve the problem?

3. For the above-level problem, what did you think about when you drew the hands on the clock to show the time?

Differentiate ⬤

Analog clocks make it easier for students to visualize the relationships among times. For example, students see how close 6:58 is to 7:00. Digital clocks are easier to read and provide a model for how to write the time. Give a digital and an analog clock to a pair of students who would benefit from extra practice. Have students show a time on the digital clock and then show the same time on the analog clock. Displaying the time on both clocks helps students to link the two representations.

Saturday Mornings

Use the clocks to answer the question.

Suzi saw one of these clocks when she woke up.

Suzi woke up before seven-thirty in the morning.

Suzi woke up after 7:00 A.M.

At what time did Suzi wake up?

Saturday Mornings

Use the clocks to answer the question.

Luke woke up just before seven o'clock in the morning.

He went to a friend's house just before 9:00 A.M.

The other clock shows the time that he had breakfast.

At what time did Luke have breakfast?

Saturday Mornings

Use the clocks to answer the question.

Wake Up   Breakfast   Soccer Game

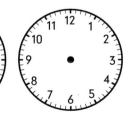

_____   _____   _____

Laney likes to sleep late on Saturday mornings. She wanted to sleep until 10:00 A.M. but woke up 19 minutes before that time. She sat down to eat breakfast just before 10:00 A.M. Though scheduled for ten-thirty in the morning, her soccer game actually began 42 minutes after she sat down for breakfast. What time did Laney's soccer game start? Write each time and then add hands to the last clock to show the time.

Pose the Question

Standards
- Uses models to identify, order, and compare numbers
- Understands relationships between measures

Overview
This problem set focuses on students posing questions. Given answers, they work backward to find associated questions. The data given suggest questions related to intervals of time.

Problem-Solving Strategies
- Work backward
- Organize information in a picture, list, table, graph, or diagram
- Guess and check or make an estimate

Materials
- *Pose the Question* (page 93; posequestion.pdf)
- *Student Response Form* (page 132; studentresponse.pdf) (optional)

Activate
1. Pose a problem, such as *Last night I talked on the phone with an old friend. We talked from 8:15 to 8:43. For how many minutes did we talk?* (28 minutes) Write the times for students to see and ask them to talk about the problem in pairs. Then, have them share their thinking with the larger group. If no one suggests using a number line to find the interval, do so yourself.

2. Explain to students that the answer to a problem is *15 minutes*, and it is their job to figure out the problem that would have this answer. Encourage students to provide multiple responses, such as *How many minutes are there until gym class? How many minutes are there between 10:00 and 10:15? How long is snack time?*

Solve
1. Distribute copies of *Pose the Question* to students. Have students work alone, in pairs, or in small groups.

2. Provide encouraging feedback, such as *I see you tried many possibilities to find the answer. I notice you are using a number line to find the amount of time.*

Debrief
1. What question did you write?
2. What helped you think of questions?
3. How could a number line help you think about these times?
4. What other answers could there be in the answer box? What questions could we ask for those answers?

Differentiate ○ □ △
Some students will be more successful if they first find all of the time intervals and then compare those numbers to the answers.

Pose the Question

Use the facts below to write a question for each answer.

Facts:

- Helen saw a magic show from 1:15 P.M. to 1:45 P.M.

- Jason saw a puppet show from 2:03 P.M. to 2:37 P.M.

| Answers: | 34 | 30 | 4 |

Use the facts below to write a question for each answer.

Facts:

- Ariel weeded the garden from 9:17 A.M. to 9:47 A.M. and raked the lawn from 10:35 A.M. to 11:06 A.M.

- Mia weeded the garden from 9:50 A.M. to 10:45 A.M.

| Answers: | 55 | 31 | 30 | 25 |

Use the facts below to write a question for each answer.

Facts:

- Ataro played the piano from 2:38 P.M. to 3:23 P.M. and then he cleaned his room from 4:36 P.M. to 5:06 P.M.

- Emily played the piano from 3:15 P.M. to 3:43 P.M. and then she cleaned her room from 4:15 P.M. to 5:10 P.M.

| Answers: | 28 | 15 | 25 | 27 |

Balance It

Standards
- Understands the basic measures of mass
- Understands the basic concept of an equality relationship

Overview
Students are shown weights to place on a balance scale so that the scale is balanced. Such a model supports the meaning of *is equal to* as being in balance, an important building block of algebraic thinking, while providing problems involving kilograms.

Problem-Solving Strategies
- Find information in a picture, list, table, graph, or diagram
- Guess and check or make an estimate

Materials
- *Balance It* (page 95; balanceit.pdf)
- balance scale (*optional*)
- sticky notes
- *Student Response Form* (page 132; studentresponse.pdf) (*optional*)

Activate
1. Ask students to stand and raise their arms out as if they were a balance scale. Ask them to imagine that a rock has been placed on their right hand and a feather on their left hand. Have them show how the scale would look. (*The left hand should be higher.*)

2. Have students show what the scale would look like if there were a whale on their left hand and a goldfish on their right. (*The left hand should be lower.*)

3. Tell students to raise both arms out until they are parallel to the floor. Ask them what it means when there is something on each side and the scale looks like this. (*The weight is equal.*)

Solve
1. Distribute copies of *Balance It* to students and make sure students understand that their task is to place each weight on the scale so the scale will be balanced. Have them discuss how they can represent the weights on the scale, e.g., that they can draw a square and write the number in it. Have students work alone, in pairs, or in small groups.

2. Listen to students' conversations. Do students' words suggest that they have a clear understanding of what it means for a scale to be balanced? Do they mention equality?

Debrief
1. How can making an incorrect guess help you to make a better guess in the future?

2. What equation can we write to show what is on the scale?

Differentiate ⬤ ◼
Guessing and checking will be easier for some students if they write the numbers on sticky notes and move them back and forth in search of an arrangement with the same sum on each side.

Balance It ○

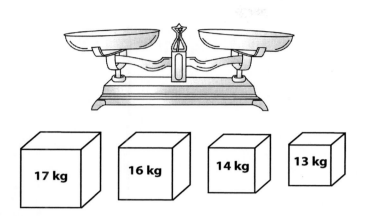

Use all of the weights.

Show the weights on the scale so that it is balanced.

How many kilograms will be on each side of the scale?

Balance It ▢

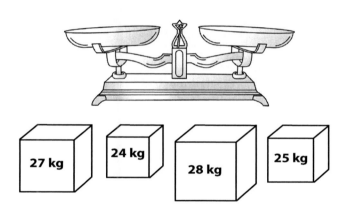

Use all of the weights.

Show the weights on the scale so that it is balanced.

How many kilograms will be on each side of the scale?

Balance It △

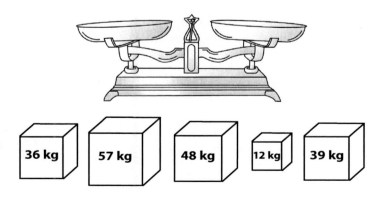

Use all of the weights.

Show the weights on the scale so that it is balanced.

How many kilograms will be on each side of the scale?

What Does It Hold?

Standards
- Determines the effects of addition on size and order of numbers
- Understands the basic measures of volume
- Knows approximate size of basic standard units and relationships among them

Overview
These problems require students to use addition to solve problems involving liters. Problem data refer to real objects and reinforces students' awareness of the capacity of common containers.

Problem-Solving Strategies
- Guess and check or make an estimate
- Work backward
- Count, compute, or write an equation

Materials
- *What Does It Hold?* (page 97; whathold.pdf)
- liter container
- *Student Response Form* (page 132; studentresponse.pdf) (*optional*)

Activate
1. Show a liter container and ask students what might be measured with this container. Allow several students to respond. Then, ask them how much liquid they think the sink will hold.

2. If there is a sink in or near your classroom, have students estimate how many liters of water it would hold. Once estimates are identified, measure to check. Then, ask whether they think a kitchen sink would hold more or less water than this sink. Ask about a bathroom sink or a laundry sink as well, and have students explain their thinking.

Solve
1. Distribute copies of *What Does It Hold?* to students. Have students work alone, in pairs, or in small groups.

2. Observe students as they work. Do they compute mentally? Do they use efficient paper and pencil strategies?

Debrief
1. What strategies did you use to solve the problem?

2. How can checking a guess help you to make the next guess?

3. How did you organize your work?

Differentiate ▢ △
The on- and above-level are more challenging since no possible choices are provided. Some students' number sense will help them to make a first guess that is quite close. Other students may lack this ability and become frustrated with the number of guesses they need to make. Ask those students refocusing questions, such as *Is the sum too big or too small? What does that tell you about the first number you guessed?*

What Does It Hol ?

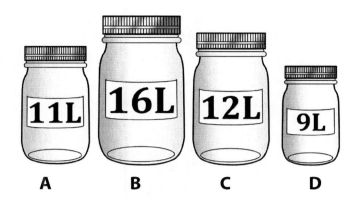

A   B   C   D

Daryl filled two of these jars with water.

In all, the jars held 25 liters of water.

Which jars did he fill?

What Does It Hold?

A large sink holds 5 more liters than a small sink.

Together they hold 21 liters of water.

How many liters of water does the large sink hold?

What Does It Hold?

A large bucket holds 4 more liters than a medium bucket.

The medium bucket holds 3 more liters than a small bucket.

In all, the buckets hold 25 liters of water.

How many more liters of water does the large bucket hold than the small one?

Moving Along

Standards

- Knows approximate size of basic standard units and relationships among them
- Uses specific strategies to estimate quantities

Overview

Students must estimate measurements to complete the scenarios about the measurements of real-world items.

Problem-Solving Strategies

- Find information in a picture, list, table, graph, or diagram
- Guess and check or make an estimate

Materials

- *Moving Along* (page 99; moving.pdf)
- meter sticks
- familiar products in boxes with the weight identified in grams
- *Student Response Form* (page 132; studentresponse.pdf) *(optional)*

Activate

1. Display the following problem:

   Use each number from the box only once to complete the story.

27	2	1

   Jack has a dog named Goldie. Jack takes Goldie for a long walk each morning. They walk about _____ kilometers. Goldie weighs _____ kilograms and drinks a little more than _____ liter of water each day. (2, 27, 1)

2. Have students discuss their strategies for filling in the numbers.

Solve

1. Distribute copies of *Moving Along* to students. Have students work in pairs so they can talk about their choices.

2. Listen to students as they work. How do they explain their choices? Do they refer to benchmarks to help them? Which units do they choose first?

3. Ask clarifying and refocusing questions, such as *What do you know about grams? Why did you place this number here?*

Debrief

1. How did you decide where to place the numbers?

2. What number did you find easiest to place? Why?

3. What blanks did you find more challenging to fill? Why?

Differentiate ⬤ ▢ △ ☆

Some students are in the habit of filling in blanks in order. Encourage them to fill in the measures that make the most sense first. Have meter sticks and products available as references.

Use each number from the box only once to complete the story so it makes sense.

1	425	90	10

Nancy is _____ years old and plays soccer. Her school's soccer field is about

_____ meters long. The soccer ball weighs about _____ grams. Practice lasts

for _____ hour every day after school.

Use each number from the box only once to complete the story so it makes sense.

15	1	132	318

Janel is about _____ centimeters tall and rides her bicycle to school. Her bicycle

has a place to keep her water bottle which holds _____ liter of water. Janel's

helmet weighs _____ grams. It takes Janel about _____ minutes to bike

to school.

Use each number from the box only once to complete the story so it makes sense.

2	600	28	58	83

Mr. Thomas weighs about _____ kilograms and likes to swim. He swims

_____ kilometers every day. He swims for just less than an hour, or _____

minutes. He swims in a pool that is about _____ meters long and holds about

_____ liters of water.

Classroom Data

Standards

- Understands that data represent specific pieces of information about real-world objects or activities
- Reads and interprets simple graphs

Overview

These problems require students to find information in a picture graph to answer questions about the data displayed.

Problem-Solving Strategies

- Count, compute, or write an equation
- Find information in a picture, list, table, graph, or diagram

Materials

- *Classroom Data* (page 101; classroom.pdf)
- *Sports Picture Graph* (sportsgraph.pdf)
- *Student Response Form* (page 132; studentresponse.pdf) *(optional)*

Activate

1. Display the *Sports Picture Graph* for students. Ask questions to make sure they understand the meaning of the graph, such as *What is this graph about? What does each stick figure stand for? Could students choose more than one category?*

2. Ask specific questions about the data, such as *How many students chose soccer? How many more students chose soccer than football? Which sport was chosen the greatest number of times?*

Solve

1. Distribute copies of *Classroom Data* to students. Have students work alone or in pairs.

2. Ask engaging questions such as, *Which season is your favorite? Which of these books have you read? Have you ever donated something to charity?*

3. Observe students as they work. Do they skip count as they count the symbols in a row? Do they use multiplication skills to find the totals? This may provide some insight on their problem-solving strategies.

Debrief

1. How did you find the answer?

2. How could you find the total in a different way?

3. What is a question you can answer by just looking at the picture graph?

4. How would you show an odd number of selections if each symbol represents an even number?

Differentiate ◯

Students who rely on skip counting or addition to find the totals may find it easier to record each number they say on the related symbol. Recording the numbers will help them keep track of the process as well as remind them of the last number identified.

Classroom Data

Favorite Seasons

= 2 students

Season	Votes
fall	😊 😊 😊
winter	😊 😊
spring	😊 😊 😊 😊
summer	😊 😊 😊 😊

Vito's class made a picture graph about their favorite seasons.

How many students chose summer as their favorite season?

Books Read

= 3 students

Book Title	Number of Students
Anansi the Spider	📗 📗 📗 📗 📗 📗
The Best Older Sister	📗 📗 📗 📗
Magic School Bus	📗 📗 📗 📗 📗 📗 📗 📗 📗
The Spy on Third Base	📗 📗 📗 📗 📗

Lacey's class made a picture graph to show how many students read each of these books.

How many fewer students read *The Spy on Third Base* than *Magic School Bus?*

Canned Food Collected

Week	Number of Cans
week 1	🥫 🥫 🥫 🥫
week 2	🥫 🥫 🥫 🥫 🥫
week 3	🥫 🥫 🥫 🥫 🥫 🥫
week 4	🥫 🥫 🥫 🥫 🥫 🥫 🥫

Noah's class collected cans of food to give to the food pantry. The picture graph shows the number of cans they collected each week.

The class collected 24 cans in week 2. How many cans does 🥫 represent?

How many more cans of food did the students collect in week 4 than week 1?

Measure It

Standards

- Understands that data represent specific pieces of information about real-world objects or activities
- Organizes and displays data in simple graphs

Overview

Students take measurements to the nearest half or quarter inch and record the measurements on a line plot. Then, they answer questions about their data.

Problem-Solving Strategies

- Count, compute, or write an equation
- Organize information in a picture, list, table, graph, or diagram

Materials

- *Measure It* (page 103; measureit.pdf)
- *Student Response Form* (page 132; studentresponse.pdf) *(optional)*

Activate

1. Ask students how they measure something to the nearest quarter inch. Make a list of students' ideas.

2. Display a line less than 12 inches long, and have students watch you measure it to the nearest half inch. Tell students that you are going to make a mistake measuring the line. Ask them to watch closely so they can tell what you do wrong. Demonstrate various errors such as not aligning the left side of the ruler to zero, rounding down to the nearest inch when the measure is closest to the half inch, and rounding up when the measure is closer to the previous half inch.

3. Add the errors to the list of students' ideas.

Solve

1. Distribute copies of *Measure It* to students. Have students work in pairs so that one student can measure while the other checks the accuracy of the measurement. Students in the pairs then reverse roles to repeat the process, thus reinforcing the practice of measuring more than once.

2. As students work, ask questions such as *How do you decide which is the closest half (quarter) inch?*

Debrief

1. How did making a line plot help you to answer the questions?

2. What answer did you find? Did anyone find a different answer?

3. Can both (all) of these answers be correct?

4. Can any number be correct?

Differentiate ◯ ▢ △ ☆

Some students may need to practice the appropriate way to take a measurement of another person. Have them practice asking someone politely if they may take a measurement and how to be respectful while measuring.

Measure It

Measure the length of five children's shoes to the nearest $\frac{1}{2}$ inch.

Make an X above the number of inches to show each measurement.

6	$6\frac{1}{2}$	7	$7\frac{1}{2}$	8	$8\frac{1}{2}$	9	$9\frac{1}{2}$	10	$10\frac{1}{2}$

Number of Inches

How many children did you measure with a shoe shorter than nine inches?

Measure It

To the nearest $\frac{1}{4}$ inch, measure the length of six students' pinky fingers.

Make an X above the number of inches to show each measurement.

0	$\frac{1}{4}$	$\frac{1}{2}$	$\frac{3}{4}$	1	$1\frac{1}{4}$	$1\frac{1}{2}$	$1\frac{3}{4}$	2	$2\frac{1}{4}$	$2\frac{1}{2}$	$2\frac{3}{4}$	3

Number of Inches

How many pinky fingers did you measure that were one to two inches long?

Measure It

Find 10 books in your classroom that are between six inches and 10 inches long. Measure each one to the nearest $\frac{1}{4}$ inch. Make an X above the number of inches to show each measurement.

6	$6\frac{1}{4}$	$6\frac{1}{2}$	$6\frac{3}{4}$	7	$7\frac{1}{4}$	$7\frac{1}{2}$	$7\frac{3}{4}$	8	$8\frac{1}{4}$	$8\frac{1}{2}$	$8\frac{3}{4}$	9	$9\frac{1}{4}$	$9\frac{1}{2}$	$9\frac{3}{4}$	10

Number of Inches

How much longer was the longest book you measured than the shortest one?

Measurement and Data

Keeping Track

Standards
- Understands that data represent specific pieces of information about real-world objects or activities
- Reads and interprets simple bar graphs

Overview
These problems require students to find information in a bar graph to answer questions about the data displayed.

Problem-Solving Strategies
- Find information in a picture, list, table, graph, or diagram
- Count, compute, or write an equation

Materials
- *Keeping Track* (page 105; keeptrack.pdf)
- *Bar Graph Template* (bargraph.pdf)
- *Student Response Form* (page 132; studentresponse.pdf) (*optional*)

Activate
1. Choose a topic of interest for your students or pose a simple question such as *Which color is your favorite?* Identify five possible choices and list them in the first column of a two-column table. Pose the question and collect the data, one row at a time, recording the totals in the right column of the table.
2. Display the *Bar Graph Template* for students and have five volunteers, one at a time, make the corresponding bar for each total. Ask them what they had to think about to draw the bar.
3. Ask questions about the information. For example, if *favorite color* were the topic, ask which color was chosen the most, least, and how many more times one color was chosen than another.

Solve
1. Distribute copies of *Keeping Track* to students. Have students work alone or in pairs.
2. Ask clarifying and refocusing questions as students work, such as *Is that bar up to the line or between two lines? Is the bar in the middle or closer to the top or bottom?*

Debrief
1. How did you find the answer?
2. How could you find the answer in a different way?
3. What do you have to think about when the bar is between the lines?

Differentiate ○ □ △ ☆
You may wish to suggest that some students align a straight edge from the top of a bar to the numbers on the vertical axis to help them determine the number the bar represents. Some students are particularly interested in data about each other. Encourage them to choose a topic, collect the data, and display the information in a bar graph.

104 #50775—50 Leveled Math Problems, Level 3 © Shell Education

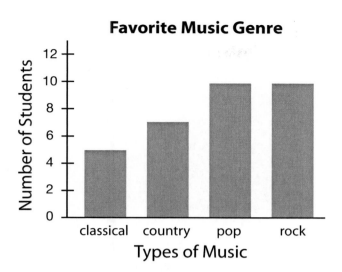

The bar graph shows the number of students who like each kind of music.

How many more students like rock than country?

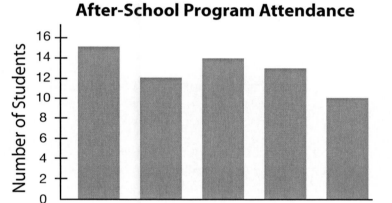

The bar graph shows the number of students in each grade who attend the after-school program.

In all, how many students in grades 1 and 2 are in the after-school program?

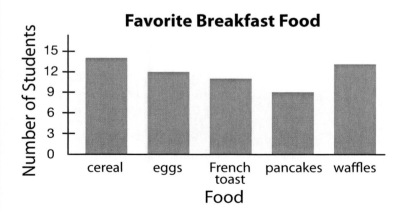

The bar graph shows students' favorite breakfast food.

How many students did not choose French toast?

Make It Yourself

Standards

- Understands the basic measures of perimeter and area
- Understands relationships between measures

Overview

Students are shown a figure on the geoboard and asked to find its area. Then, they draw a figure on the other geoboard that meets a stated criteria related to the first figure.

Problem-Solving Strategies

- Act it out or use manipulatives
- Count, compute, or write an equation
- Guess and check or make an estimate

Materials

- *Make It Yourself* (page 107; makeyourself.pdf)
- geoboards and elastic bands (or string)
- dot paper *(optional)*
- *Student Response Form* (page 132; studentresponse.pdf) *(optional)*

Activate

1. Use an elastic band to indicate a small square on a geoboard and tell students that this is one square unit. Point to the length of one side and tell students that it is one unit long. (Alternatively, you could use dot paper for this.)

2. Place an elastic band to indicate a 2 × 4 rectangle and have students create the figure on their boards. Ask *How many square units are inside this rectangle? How do you know? What are we measuring when we find the number of square units inside a figure?*

3. Trace your finger along the outline of the figure and ask *What is the perimeter of this figure?*

4. Repeat with a figure that is composed of multiple rectangles or squares, such as the figure below.

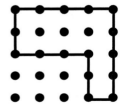

Solve

1. Distribute copies of *Make It Yourself* to students. Have students work alone, in pairs, or in small groups.

2. Listen to students as they work. Do they refer to square units appropriately?

Debrief

1. How did you find the area? Did anyone use a different strategy?

2. What strategies did you use to decide on the figure to draw?

Differentiate ⬤ ▢ △ ☆

Some students may prefer to work with the geoboards while others may prefer working with pencils on dot paper. Allow them to follow their preferences unless their choice leads to frustration, for example, when guessing and checking on paper requires too much erasing.

Make It Yourself

What is the area of this figure?

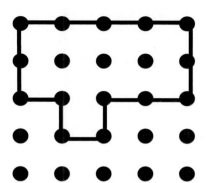

Draw a square with the same area.

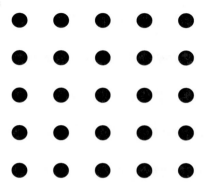

Make It Yourself

What is the area of this figure?

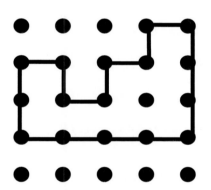

Draw a different figure with the same perimeter.

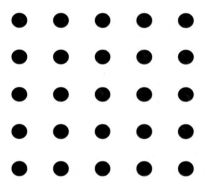

Make It Yourself

What is the area of this figure?

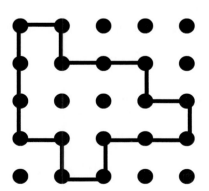

Draw a figure with the same perimeter but a smaller area.

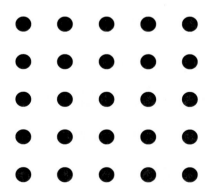

Which One?

Standards

- Multiplies whole numbers
- Understands the basic measures of area

Overview

These problems require students to find the area of objects given their length measures. The goal is to find the object that matches the given criteria.

Problem-Solving Strategies

- Find information in a picture, list, table, graph, or diagram
- Count, compute, or write an equation
- Guess and check or make an estimate

Materials

- *Which One?* (page 109; whichone.pdf)
- graph paper
- *Student Response Form* (page 132; studentresponse.pdf) (*optional*)

Activate

1. Ask students what they know about *area*. Allow several students to respond.

2. If no one refers to finding the area of a rectangle, ask *Who can tell me something about finding the area of a rectangle?*

3. Draw a rectangle and label the length *6 ft.* and the width *2 ft.* Point to the sides that are not labeled and ask students to identify the lengths. Then, ask students what the area of this rectangle is. Make sure students respond *12 square feet* rather than just *12*. If necessary, remind them that a measurement requires a number and a unit.

Solve

1. Distribute copies of *Which One?* to students. Have students work alone, in pairs, or in small groups.

2. Observe students as they work. Do they readily compute the areas of the figures? Does their number sense allow them to eliminate some possibilities without finding their areas?

Debrief

1. What strategies did you use to find the answer?

2. Did you always use the clues in order? Why or why not?

3. Did you check each area? Why or why not?

Differentiate ◯ ▢ △ ☆

Distribute graph paper to students who prefer a more concrete approach to finding areas. When needed, they can draw the figure and count the squares. Make sure students who rely on the formula remember why it works. Ask *What could you show someone who does not understand why the formula for finding the area of a rectangle works?*

Which One? ○

Harry bought a rug that has an area of 24 square feet. Which rug did Harry buy?

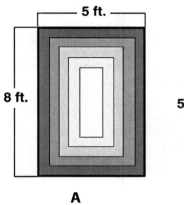
A

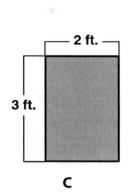
B

C

D

Which One? □

The area of Sally's bedroom is greater than 72 square feet. The area of the room is not 80 square feet. Which drawing below represents Sally's bedroom?

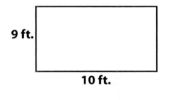 9 ft. 10 ft.
A

8 ft. 9 ft.
B

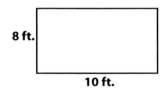 8 ft. 10 ft.
C

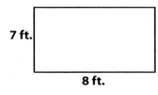 7 ft. 8 ft.
D

Which One? △

The area of Madelyn's picture is 124 square inches greater than the area of Ben's picture. Which pictures did Madelyn and Ben draw?

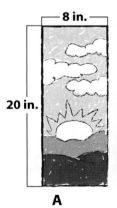
A

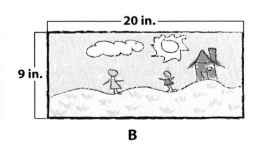
B

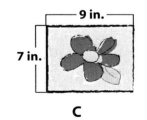
C

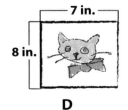
D

Hobbies

Standards

- Multiplies whole numbers
- Understands the properties of and the relationships among addition and multiplication
- Understands the basic measures of area

Overview

In these problems students make connections between the distributive property and area.

Problem-Solving Strategies

- Count, compute, or write an equation
- Find information in a picture, list, table, graph, or diagram
- Organize information in a picture, list, table, graph, or diagram

Materials

- *Hobbies* (page 111; hobbies.pdf)
- graph paper
- *Student Response Form* (page 132; studentresponse.pdf) *(optional)*

Activate

1. Share the following scenario with students: *I used tiles to decorate a box. Each tile has an area of one square inch. I made three rows of four yellow tiles. Right beside them I made three rows of two blue tiles.* Distribute graph paper to students. Read the scenario again and ask students to represent the data on the graph paper. Have students check their representations with a neighbor.

2. Ask *What is the area of the yellow tiles? (12 square inches) What is the area of the blue tiles? (6 square inches) What is the area of all the tiles? (18 square inches) What equation can we write to show how we found the area of all the tiles? Is there another way we could find the area? Is there another equation we could write? How does this area problem relate to the distributive property?*

Solve

1. Distribute copies of *Hobbies* to students. Have students work alone, in pairs, or in small groups. Make graph paper available to students.

2. Ask refocusing and clarifying questions as students work, such as *What are you trying to find? Is there another way you could find the area? What picture could you draw to help you with this problem?*

Debrief

1. How do diagrams help you solve problems?

2. How does the problem relate to the distributive property?

Differentiate

You may wish to direct some students to show the drawings on graph paper so that they can see the tiling of the area.

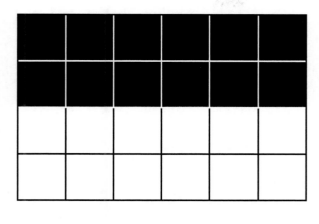

Jed made this design with small tiles.

The area of each tile is one square inch.

What is the area of the black tiles?

What is the total area of these tiles?

Corinne is knitting a scarf with blue and green stripes. The scarf is 8 inches wide. She knits a blue stripe that is 4 inches long. She knits a green stripe that is 5 inches long.

What is the area of the scarf so far?

Write an equation to show how you can find the area.

Jason made this drawing to solve a problem. He found the correct answer of 320 square feet.

Write a problem Jason might have solved.

Write two equations Jason might have used to find the answer.

10 ft.   **30 ft.**

8 ft.

All Around the Garden

Standards

- Understands the basic measures of perimeter and area
- Understands relationships between measures
- Understands basic properties of figures

Overview

These problems focus on finding the perimeter of squares and rectangles.

Problem-Solving Strategies

- Find information in a picture, list, table, graph, or diagram
- Organize information in a picture, list, table, graph, or diagram
- Count, compute, or write an equation

Materials

- *All Around the Garden* (page 113; aroundgarden.pdf)
- *Student Response Form* (page 132; studentresponse.pdf) *(optional)*

Activate

1. Ask students what they might measure in a garden, and why. Allow several students to respond and make a list of their ideas. Be sure the list includes area and perimeter, and that students are clear on the difference between the two.

2. Display a drawing of a square with one side labeled 4 feet. Ask *What do we know about the other sides of the square? (They are also 4 feet.) How might we find the perimeter of this figure? Is there another way we could find it?*

Solve

1. Distribute copies of *All Around the Garden* to students. Have students work alone or in pairs.

2. Observe students as they work. Do they readily recognize the lengths of the other sides? Do they compute mentally or use paper and pencil techniques?

Debrief

1. How did you find the perimeter? What equation could we write to show this method?

2. What is another way to find the perimeter? What equation could we write to show this method?

3. How could a drawing help you solve a problem involving area or perimeter?

Differentiate

You may wish to label all of the length measures for some students. With all the lengths given, students can focus on the definition of perimeter and how to find it.

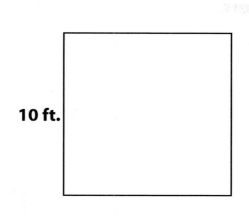

10 ft.

What is the perimeter of this square garden?

Mr. Wilson made a garden in the shape of a square. Ms. Diaz made a garden in the shape of a rectangle. How much greater is the perimeter of Mr. Wilson's garden than Ms. Diaz's garden?

5 ft.

5 ft.

3 ft.

Loretta and Bill made a new garden for tomatoes this year. It is 6 feet wide. The area of this garden is 48 square feet. What is the perimeter of this garden?

Around the Garden (side tabs)

Many Measures

Standards

- Understands the basic measures of perimeter and area
- Understands relationships between measures

Overview

Students list the possible dimensions of a figure with a given area. Once they have made a list, they answer a question involving perimeter.

Problem-Solving Strategies

- Act it out or use manipulatives
- Count, compute, or write an equation
- Organize information in a picture, list, table, graph, or diagram
- Work backward

Materials

- *Many Measures* (page 115; manymeasures.pdf)
- graph paper
- *Student Response Form* (page 132; studentresponse.pdf) *(optional)*

Activate

1. Direct students to use graph paper to show how many different rectangles they can make with an area of 20 square feet. Have students share their ideas. Note that some students may suggest that a 5×4 rectangle is the same as a 4×5 rectangle, while others may believe they are different. Either interpretation is fine; just make sure the class makes a joint decision.

2. Ask *What are the perimeters of these rectangles? Which rectangle has the longest perimeter? Which figure has the shortest perimeter?*

Solve

1. Distribute copies of *Many Measures* to students. Have students work alone, in pairs, or in small groups.

2. Ask clarifying and refocusing questions as appropriate. For example, *What makes you think you have found all of the rectangles? How might thinking about products help you? Is a square also a rectangle?*

Debrief

1. What helped you to find the different rectangles with that area?

2. How did you organize your list? What is another way to organize the list?

3. What did you notice about the way the perimeters change?

Differentiate ⬤ ◻ △ ☆

Some students may succeed by drawing different rectangles but not always record what they find in a list. Encourage those students to work with a partner, alternating the responsibility of being the recorder. You may wish to pose the following exit-card task: *Which rectangle with an area of 30 square units will have the greatest perimeter?* *(1 × 30)*

Many Measures ◯

What rectangles can you make with an area of 12 square feet? List the lengths and widths.

Which rectangle has the perimeter of 16 feet?

Many Measures ▢

What rectangles can you make with an area of 18 square feet? List the lengths and widths.

Which rectangle has the shortest perimeter? What is that perimeter?

Many Measures ◁

You make a rectangle with an area of 24 square feet. What could the rectangle's perimeter be?

Your friend makes a rectangle with an area of 16 square feet. This rectangle has a perimeter that is the same as the perimeter of your rectangle.

What are the side lengths of the rectangle you made?

What are the side lengths of the rectangle your friend made?

Tangram Shapes

Standards

- Knows basic geometric language for describing and naming shapes
- Predicts and verifies the effects of combining, subdividing, and changing basic shapes

Overview

Students arrange a set of shapes to make a specified shape such as a rectangle or a quadrilateral.

Problem-Solving Strategies

- Act it out or use manipulatives
- Guess and check or make an estimate
- Use logical reasoning

Materials

- *Tangram Shapes* (page 117; tangram.pdf)
- tangram sets
- *Student Response Form* (page 132; studentresponse.pdf) *(optional)*

Activate

1. Distribute the two smallest triangles and the square from a tangram set to students. Challenge students to use both of the two smallest triangles to make different polygons. Ask students if they can make a square. Once students have formed one, ask them what strategies they used to make it.

2. Challenge students to use these two pieces to make a bigger triangle and then a parallelogram. Again, have students describe the strategies they used to find the shapes. Did they picture the shape in their minds? Did they guess and check where to put the pieces?

Solve

1. Distribute copies of *Tangram Shapes* to students. Have students work alone or in pairs. Have students outline the shapes that make the figure on another sheet of paper.

2. Ask clarifying and refocusing questions as students work. For example, *Which pieces are you supposed to use? What shape are you trying to make? How would you describe this shape?*

Debrief

1. Where did you place the pieces? Could they be placed differently?

2. What strategies did you use to find the shape?

3. What quadrilaterals did you make?

Differentiate ◯ ▢ △ ☆

Some students may experience frustration when trying to outline the shapes. Suggest that these students work in pairs with one student holding down the pieces while the other student outlines the shape.

angram Shapes

Use the square and two small triangles.

How can you use all of these shapes to make a rectangle? Draw the rectangle.

How can you use all of these shapes to make a triangle? Draw the triangle.

angram Shapes

Use the three triangles and the square.

How can you use all of these shapes to make a rectangle? Draw the rectangle.

How can you use all of these shapes to make a parallelogram? Draw the parallelogram.

angram Shapes

Use the three triangles.

Make four different quadrilaterals.

Draw the quadrilaterals.

What is the name of each shape?

Tell Me More

Standards

- Knows basic geometric language for describing and naming shapes
- Understands basic properties of figures

Overview

Students identify properties of a given shape. Then, they classify the shapes.

Problem-Solving Strategy

Use logical reasoning

Materials

- *Tell Me More* (page 119; tellmemore.pdf)
- shapes that are squares, rectangles, parallelograms, and rhombuses
- magazines or picture books (*optional*)
- *Shapes Shapes Shapes* by Tana Hoban (*optional*)
- *Student Response Form* (page 132; studentresponse.pdf) (*optional*)

Activate

1. Tell students *I am a triangle. What else do you know about me?*

2. Once students have identified a triangle as a plane shape, a polygon, and as having three sides and three angles, ask them where they see triangles. Though students are likely to mention common classroom items such as the green triangle in the pattern block set, encourage students to think about examples outside of the classroom. Record their ideas.

3. If time allows, show students a book, such as *Shapes Shapes Shapes* by Tana Hoban, that shows photographs of many shapes in the real world.

Solve

1. Distribute copies of *Tell Me More* to students. Have students work alone or in pairs.

2. Provide encouraging feedback as students work, such as *I see you have made a list of several ideas. You are listening carefully to each other's ideas.*

Debrief

1. What else do you know about the shape?

2. What helped you to identify places where you see the shape?

Differentiate ⬤

Some students will find it easier to describe attributes of a shape if they can hold examples of it. Provide magazines or picture books to students who are unable to think of where they see the shape in the real world. Looking at the pictures may help students identify other examples.

Tell Me More

I am a square.

What are two things you know about me?

Where do you see my shape?

Tell Me More

I am a rectangle.

What are three things you know about me?

Where do you see my shape?

Can I be a square?

Tell Me More

I am a parallelogram.

What do you know about me?

Where do you see my shape?

Can I be a rhombus?

Name the Shape

Standard

- Knows basic geometric language for describing and naming shapes
- Understands basic properties of figures

Overview

Students are given rows and columns of shapes. They use clues to identify the correct shape.

Problem-Solving Strategies

- Find information in a picture, list, table, graph, or diagram
- Use logical reasoning

Materials

- *Name the Shape* (page 121; nameshape.pdf)
- *Shape Matrix* (shapematrix.pdf)
- *Student Response Form* (page 132; studentresponse.pdf) *(optional)*

Activate

1. Display the shape matrix below for students.

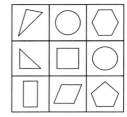

   Have students name the shapes as you point to them randomly. For the rhombus (parallelogram or quadrilateral), square (rectangle or quadrilateral), and rectangle (parallelogram or quadrilateral), ask them what else this shape could be called.

2. Tell students that you are thinking of one of these shapes. They are to listen to clues so that they can name the shape. Read the following clues, one at a time: *It has four sides.* (It is the square, rhombus, or rectangle.) *Not all of its sides are the same length.* (It is a rectangle.) After each clue, ask, *What do we know about the shape now? What shapes could we cross out?* Repeat with the following clues: *It is in the first row. It is in the second row.* (It is the circle.)

Solve

1. Distribute copies of *Name the Shape* to students. Have students work alone or in pairs.

2. Observe students as they work. Do they confuse rows and columns? Do they recognize shapes readily?

Debrief

1. Which shape did you name? Why do you think it is the shape?

2. What other names can we give this shape?

Differentiate ◯ ▢ △ ☆

To prepare for further instruction have students complete an exit card with the following task: *Name three different shapes that are quadrilaterals.*

Name the Shape ◯

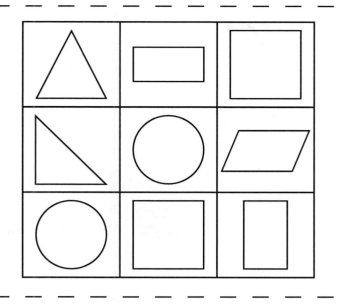

I am in a row with a square.

I am next to a triangle.

I am a _____.

Name the Shape ▢

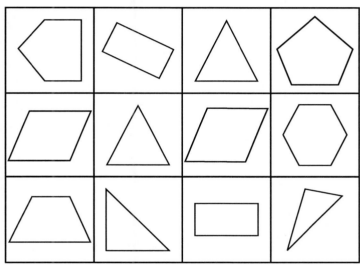

All of my sides have the same length.

I am in a column with a rectangle.

I am not a triangle.

I am not a rectangle.

I am a _____.

What are two other names I can be called?

I am a _____ or a _____.

Name the Shape △

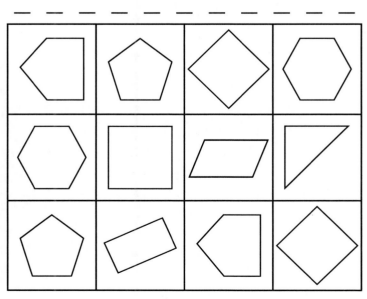

There are exactly two quadrilaterals in my row.

There are exactly two quadrilaterals in my column.

There is not a triangle in my row.

I am not a pentagon.

I am a _____ .

What are two other names I can be called?

Table Shapes

Standards

- Multiplies and divides whole numbers
- Knows basic geometric language for describing and naming shapes

Overview

Students identify the number of sides in a square, pentagon, or hexagon to determine the number of chairs that could fit at a table of that shape. They then use multiplication and division to find the total number of chairs given the number of tables, or the number of tables given the number of chairs.

Problem-Solving Strategies

- Count, compute, or write an equation
- Organize information in a picture, list, table, graph, or diagram
- Guess and check or make an estimate

Materials

- *Table Shapes* (page 123; tableshapes.pdf)
- *Student Response Form* (page 132; studentresponse.pdf) (*optional*)

Activate

1. Ask students what polygons they can name, and how many sides they have. Have students work with a partner to make a list and then share their ideas as you record them.

2. Tell students to draw a picture of a pentagon and to imagine that there is a table with a top that is this shape. Ask *If there is a chair at each side of this table, how many chairs would there be?* (5) *How many chairs would there be at three tables like this one?* (15)

3. Display the following problem: *There are 36 chairs at tables. Each table is the same shape. There is one chair at each side of the table. There are six tables. What shape are the tables?* (hexagons)

Solve

1. Distribute copies of *Table Shapes* to students. Have students work alone or in pairs.

2. Ask refocusing or clarifying questions as students work, such as *What does this number stand for? How are you organizing your information? What do you know about the tables? How might a guess help you find the number of tables?*

Debrief

1. How did you find the answer?

2. Which strategy did you find the most helpful?

3. What equation could you write to represent this situation?

Differentiate ⬤ ▲

Some students may need to sit near the list you made of the shapes so that they can easily identify the number of sides for each polygon. Some students may have more success with the above-level problem if you encourage them to begin by making a guess for one of the table shapes.

able Shapes

There are two tables with a pentagon shape.

There is one table with a square shape.

There is a chair on each side of each table.

How many chairs are there?

---

a le Shapes

There are three tables with a square shape.

There are some tables with a hexagon shape.

There is a chair on each side of each table.

There are a total of 36 chairs.

How many tables are there with a hexagon shape?

Table Shapes

There are some tables with a pentagon shape and some tables with a hexagon shape.

There are three more tables with a pentagon shape than with a hexagon shape.

There is a chair on each side of each table.

There are 59 chairs.

How many tables of each shape are there?

Shape Sentences

Standards

- Knows basic geometric language for describing and naming shapes
- Understands basic properties of figures

Overview

Students are shown four shapes and asked to complete one or more statements about the figures. All of the sentences begin with *all*, *some*, or *none*.

Problem-Solving Strategies

- Find information in a picture, list, table, graph, or diagram
- Use logical reasoning

Materials

- *Shape Sentences* (page 125; shapesentences.pdf)
- *Student Response Form* (page 132; studentresponse.pdf) (*optional*)

Activate

1. Have four students stand in front of the class. Ask students how they might complete this sentence: *All of these students….* Collect responses and then ask students what has to be true when we begin a sentence with *All of these….*

2. This time ask students to complete the sentence *Some of these students….* Again, ask students to share their thinking.

3. Present the task of finishing the sentence *None of these students….*

4. Tell students that they are going to consider the same task, but the sentences will be about shapes rather than students.

Solve

1. Distribute copies of *Shape Sentences* to students. Have students work alone or in pairs.

2. Ask clarifying and refocusing questions as students work, such as *How would you describe this shape? How is this shape different from that one? What do these two shapes have in common?*

Debrief

1. What strategies did you use to find the answers?

2. What is a different possible answer?

3. What do you look for if you are writing a sentence that begins with *Some of these shapes...*?

4. Do you think it is easier to complete a sentence that starts with *all*, *some*, or *none*? Why?

Differentiate ◯ ▢ △ ☆

To better prepare future instruction have students complete this exit card task: *Draw four different shapes that fit the following sentence: All of these shapes are quadrilaterals.*

Shape Sentences

How could you finish this sentence?

All of these shapes are _____.

Shape Sentences

How could you finish these sentences?

Some of these shapes are _____.

None of these shapes are _____.

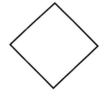

Shape Sentences

How could you finish these sentences?

All of these shapes are _____.

Some of these shapes are _____.

None of these shapes are _____.

They Belong Together

Standards

- Knows basic geometric language for describing and naming shapes
- Understands basic properties of figures

Overview

Students are shown four shapes and asked to focus on the similarities and differences among them.

Problem-Solving Strategies

- Find information in a picture, list, table, graph, or diagram
- Use logical reasoning

Materials

- *They Belong Together* (page 127; theybelong.pdf)
- *Student Response Form* (page 132; studentresponse.pdf) (*optional*)

Activate

1. Display the shapes below and ask students which shape they think is different, and why. Ask students to find another reason this shape does not belong with the group. Then, ask them if they can find another shape that does not belong. Provide time for students to think alone before talking with a partner.

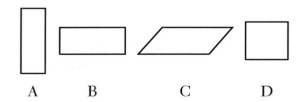

    A   B   C   D

2. Have the partners share their thinking with the larger group. Encourage a variety of responses. Some students may suggest shape *A* since it is the only figure with a vertical length greater than its horizontal length. Other students might suggest shape *D* since it is the only shape with four sides that have the same length. Shape *C* could also be chosen, as it is the only figure without right angles.

Solve

1. Distribute copies of *They Belong Together* to students. Have students work alone or in pairs.

2. Ask clarifying and refocusing questions as students work. For example, *What else can you tell me about this shape? Tell me more about why you think this shape is different.*

Debrief

1. Which shape do you think does not belong? Why?

2. What is a different reason or a different shape that could not belong?

Differentiate ⬤

Some students may need support to recognize similarities and differences. You may want to suggest that these students talk about each shape with a partner, recording their ideas as they do so. Students can then use these lists to help them identify ways in which a shape might be different.

They Belong together

Which shape does not belong with the others? Why?

A   B   C   D

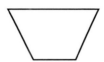

They Belong together

Which shape does not belong with the others? Why?

A   B   C   D

They Belong Together

Which shape does not belong with the others? Why?

  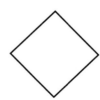 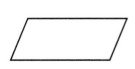

A   B   C   D

Draw Me

Standards

- Knows basic geometric language for describing and naming shapes
- Understands basic properties of figures

Overview

Students draw a shape to meet a given description of its properties.

Problem-Solving Strategies

- Organize information in a picture, list, table, graph, or diagram
- Use logical reasoning

Materials

- *Draw Me* (page 129; drawme.pdf)
- examples of quadrilaterals
- math dictionaries
- *Student Response Form* (page 132; studentresponse.pdf) *(optional)*

Activate

1. Ask students what they can tell you about a polygon. Give several students the opportunity to respond.

2. Ask *What do you know about a polygon that is a quadrilateral?*

3. Tell students that you are going to describe a polygon for them to draw. Write the following directions as you say to them: *I am a quadrilateral. I am a rectangle.*

4. Encourage students to share their images. Ask *What is the same about these figures? What is different?*

5. Many students do not recognize a square as a rectangle. If no one drew a square, draw one and ask students if it matches the description.

Solve

1. Distribute copies of *Draw Me* to students. Have students work alone, in pairs, or in small groups.

2. Observe students as they work. Do they appear to form images of shapes in their minds or draw shapes to check against the problem criteria? This will provide some insight into their problem-solving processes.

3. Listen to students as they talk about the problems. How do they describe the shapes and their properties?

Debrief

1. How did you decide what shape to draw? Did anyone draw a different shape?

2. Which clue did you find most helpful? Why?

3. What strategies would you suggest someone use when solving these problems?

Differentiate

Assign students to work in pairs so that they can discuss the meanings of the geometric terms. Provide a few students with samples of quadrilaterals from which they can choose the one that matches the description. Some students might be more successful with ready access to a math dictionary.

Draw Me

I am a quadrilateral.

My opposite sides have the same length.

I am not a square.

How might I look?

Draw Me

I am a quadrilateral.

Only two of my sides have the same length.

How might I look?

Show two different ways.

Draw Me

I am a quadrilateral.

I am not a rectangle.

I am not a parallelogram.

How might I look?

Show three different ways.

Parts of Shapes

Standards

- Understands the basic measures of area
- Knows basic geometric language for describing and naming shapes
- Predicts and verifies the effects of combining, subdividing, and changing basic shapes

Overview

These problems present a larger shape with a smaller shape inside of it that is shaded. Students continue to partition the larger shape and identify the part of the larger shape that is shaded.

Problem-Solving Strategies

- Find information in a picture, list, table, graph, or diagram
- Count, compute, or write an equation

Materials

- *Parts of Shapes* (page 131; partsshapes.pdf)
- scissors
- *Student Response Form* (page 132; studentresponse.pdf) *(optional)*

Activate

1. Sketch the bar below and ask students what part of the area of the large rectangle is shaded, and to explain their thinking.

   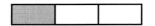

2. This time show students a rectangle with only a shaded part and ask, *What part of the area of this rectangle do you think is shaded? Why do you think so?*

3. Provide time for students to talk to a partner and then share their ideas with the larger group. Then invite a student to partition the larger rectangle to show that $\frac{1}{4}$ of the area is shaded.

Solve

1. Distribute copies of *Parts of Shapes* to students. Have students work alone or in pairs.

2. Provide encouraging feedback as students work, such as *I see you are carefully partitioning the larger figure.*

Debrief

1. How did you find the portion of the figure that was shaded?

2. What did you find most challenging about the problem?

Differentiate ⬤

Provide students two copies of the problem(s). The shaded portion can be cut from one copy and placed on the other copy to determine the number that would fit in the figure.

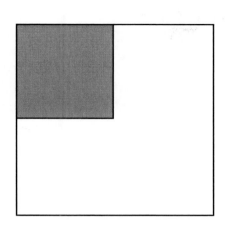

Brian drew the large square. Then he drew the small square and shaded it. How many small squares like this could Brian fit in the larger square?

What part of the area of the large square is shaded?

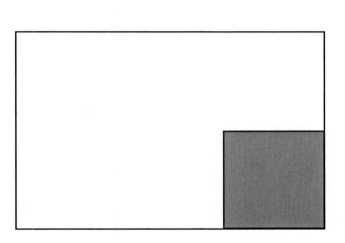

Colleen drew this rectangle. Then she drew the small square and shaded it. How many more small squares like this one could Colleen fit in the rectangle?

What part of the area of the rectangle is shaded?

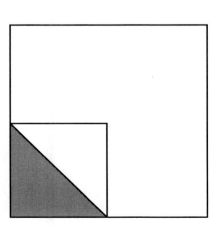

Leslie drew the large and small squares. Tom drew the triangle and shaded it. What part of the area of the large square is shaded?

Name: _____   Date: _____

Student Response Form

Problem:

(glue your problem here)

My Work and Illustrations:
(picture, table, list, graph)

My Solution:

My Explanation:

_____

_____

_____

Individual Observation Form

Name: _____ Date: _____

Shows Understanding (Check all that apply.)

☐ Makes representations or notes to understand more fully.

☐ Talks with a peer to understand more fully.

☐ Asks teacher questions to understand more fully.

☐ Interprets problem correctly.

Applies Strategies (Check all that apply.)

☐ Demonstrates use of an appropriate strategy.

☐ Tries an alternative approach when first attempt is unsuccessful.

☐ Uses a strategy appropriately after it is suggested by someone else.

Explains or Justifies Thinking (Check all that apply.)

☐ Communicates thinking clearly.

☐ Uses words and labels to summarize steps to solution.

☐ Provides mathematical justifications for solution or solution process.

☐ Uses correct mathematical vocabulary.

Takes It Further (Check all that apply.)

☐ Makes connections among problems.

☐ Poses new related problems.

☐ Solves a problem in more than one way.

Group Observation Form

Use this form to record scores, comments, or both.

Date: _____

Scores: 1—Beginning   2—Developing   3—Meeting   4—Exceeding

Group Members	Provides leadership/ suggestions to group	Builds on the comments of others	Communicates clearly, uses correct mathematical vocabulary, and builds on the ideas of others	Creates at least one accurate representation of the problem	Suggests/ chooses appropriate strategies					

#50775—50 Leveled Math Problems, Level 3

Record-Keeping Chart

Use this chart to record the problems that were completed. Record the name of the lesson and the date when the appropriate level was completed.

Name: _____

Lesson	● Date Completed	■ Date Completed	▲ Date Completed

Answer Key

Snack Time (page 33)

- ● 18 carrot sticks
- ■ 16 apples
- ▲ 42 almonds

Floor Tiles (page 35)

- ● 15 tiles
- ■ 55 tiles
- ▲ 30 tiles

Equal Groups (page 37)

- ● $10 \div 5 = \square$; 2 shelves
- ■ $24 \div 4 = \square$; 9 jumping jacks
- ▲ $(91 - 19) \div 9 = \square$; 8 pencils

Boxes of Cupcakes (page 39)

(Note: Any letter may be used and the order of factors in multiplication equations may be reversed.)

- ● $4 \times 6 = c$; 24 cupcakes
- ■ $8 \times b = 40$ or $40 \div 8 = b$; 5 boxes
- ▲ $8 \times b + 4 \times b = 84$; 7 boxes

Pattern Questions (page 41)

(Note: Any letter may be used and the order of factors in multiplication equations may be reversed.)

- ● 18; Questions will vary.
- ■ 28; $10 \times 4 = n$; Questions will vary.
- ▲ 48; $6 \times n = 54$ or $54 \div 6 = n$; 56; Questions will vary.

First Names (page 43)

- ● 4 letters
- ■ 7 letters
- ▲ 8 letters

Pose a Problem (page 45)

- ● $5 \times 4 = ?$; story problems will vary, but should match the equation; 20
- ■ $4 \times ? = 32$ or $32 \div 4 = ?$; story problems will vary, but should match the equation; 8
- ▲ $6 \times ? = 42$ or $42 \div 6 = ?$; story problems will vary, but should match the equation; 7

Finish the Steps (page 47)

- ● Any two of the following: 1 and 2, 2 and 7, 5 and 10, 10 and 11
- ■ Any three of the following: 2 and 6, 3 and 9, 6 and 12, 9 and 13, 18 and 14
- ▲ All of the following: 2 and 8, 3 and 12, 4 and 14, 6 and 16, 8 and 17, 12 and 18, 24 and 19

Pattern Hunt (page 49)

- ● Patterns will vary; 5, 11
- ■ Patterns will vary; 6, 21, 7
- ▲ Patterns will vary; 7, 56, 8, 7

Boxes and Boxes (page 51)

- ● 27 beads
- ■ 48 cards
- ▲ 95 books

Figure It (page 53)

- ● 6, 5
- ■ 7, 56
- ▲ 8, 36, 9

At the Fair (page 55)

- ● 30; 6 (or 3); 3 (or 6); 18
- ■ 4 (or 6); 6 (or 4); 15, 39, 9, 18
- ▲ 7 (or 4), 97, 6 (or 9); 9 (or 6); 43, 4 (or 7)

Answer Key *(cont.)*

Yard Sale (page 57)

- 3 building block kits and 1 puzzle
- ■ 1 bat and 3 mitts or 1 baseball and 3 mitts
- ▲ 3 building sets and 1 video game; 3 video games, 1 building set, and 1 art kit; or 2 video games, 2 building sets, and 1 board game

What's Going On? (page 59)

- 3; Choices will vary, but the answer should be the starting number. Explanations should include reference to the fact that dividing by 5 undoes or is the inverse of multiplying by 5.
- ■ 6; Choices will vary, but the answer should be the starting number. Explanations should include reference to the fact that multiplying by 4 undoes or is the inverse of dividing by 4 and subtracting 2 undoes or is the inverse of adding 2.
- ▲ 8; Choices will vary, but the answer should be the starting number. Explanations should include reference to the fact that dividing by 6 undoes or is the inverse of multiplying by 6 and multiplying by 3 and doubling is the same as multiplying by 6.

Number Models (page 61)

- 45, 46, 50–54
- ■ 215–224
- ▲ Jackson wrote nine more numbers.

Buildopoly (page 63)

- Nicola; 1 point
- ■ Marcel; 10 points
- ▲ Aisha; 5 points

Toy Store (page 65)

- 165 jump ropes
- ■ 423 baseballs
- ▲ 617 puzzles

Some Sums (page 67)

- 402 + 13 or 403 + 12 or 412 + 03
- ■ Any two of the following: 791 + 53; 793 + 51; 751 + 93; 753 + 91
- ▲ Any four of the following: 436 + 257; 437 + 256; 456 + 237; 457 + 236; 236 + 457; 237 + 456; 256 + 437; 257 + 436

Animal Facts (page 69)

- 260
- ■ 248
- ▲ 473

Family Trips (page 71)

- Cove Beach
- ■ Wild Mountain
- ▲ 298 miles

Make It True (page 73)

- 98 and 34 (or opposite order); 564 and 348
- ■ 195 and 147 (or opposite order); 746 and 257; 57 and 108
- ▲ 389 and 295 (or opposite order); 525 and 479; 479 and 295 (or opposite order)

Town Races (page 75)

- 40 children; 5 pairs
- ■ 120 children; 9 teams
- ▲ 10 teams

Make It Match (page 77)

- $\frac{1}{3}$
- ■ $\frac{4}{7}$
- ▲ $\frac{5}{6}$; $\frac{1}{6}$

Answer Key *(cont.)*

On the Number Line (page 79)

● [number line: 0 ... Paul ... 1]

■ [number line: 0 ... Marie ... Jason ... 1]

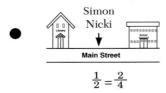

▲ [number line: 0 ... Ruby ... Liam ... DeShawn ... 1]

On the Trail (page 81)

● Mika

■ Philip

▲ Chris

Standing in Line (page 83)

● $\frac{1}{4}$

■ $\frac{2}{6}$ or $\frac{1}{3}$

▲ $\frac{5}{8}$

Who Is Where? (page 85)

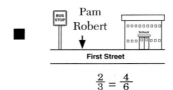

● Simon / Nicki — Main Street

$\frac{1}{2} = \frac{2}{4}$

■ Pam / Robert — First Street

$\frac{2}{3} = \frac{4}{6}$

▲ Jackie / Zak / Sophie — Winter Street

Possible statements include: $\frac{3}{4} = \frac{6}{8}$; $\frac{3}{4} > \frac{3}{8}$; $\frac{6}{8} > \frac{3}{8}$; $\frac{3}{8} < \frac{3}{4}$; $\frac{3}{8} < \frac{6}{8}$

How Much Money? (page 87)

● 18¢

■ 80¢

▲ 35¢

What Is the Number? (page 89)

● $\frac{1}{2}$

■ $\frac{3}{8}$

▲ $\frac{5}{8}$

Saturday Mornings (page 91)

● 7:14 A.M.

■ 8:32 A.M.

▲ 9:41 A.M.;
9:56 A.M.;
10:38 A.M.

Pose the Question (page 93)

Possible questions include:

● 34: How many minutes long was the puppet show?

30: How many minutes long was the magic show?

4: How many minutes longer was the puppet show than the magic show?

■ 55: For how many minutes did Mia weed the garden?

31: For how many minutes did Ariel rake the lawn?

30: For how many minutes did Ariel weed the garden?

25: How many more minutes did Mia weed than Ariel?

▲ 28: For how many minutes did Emily play the piano?

15: How many more minutes did Ataro spend playing the piano than cleaning his room?

25: How many more minutes did Emily spend cleaning her room than Ataro spent cleaning his?

27: How many more minutes did Emily spend cleaning her room than playing the piano?

Answer Key *(cont.)*

Balance It (page 95)

- ● The 17 kg and 13 kg weights should be on one side and the 16 kg and 14 kg weights on the other; 30 kg
- ■ The 28 kg and 24 kg weights should be on one side and the 27 kg and 25 kg weights on the other; 52 kg
- ▲ The 57 kg and 39 kg weights should be on one side and the 48 kg, 12 kg, and 36 kg weights on the other; 96 kg

What Does It Hold? (page 97)

- ● B and D
- ■ 13 L
- ▲ 7 L

Moving Along (page 99)

- ● 10; 90; 425; 1
- ■ 132; 1; 318; 15
- ▲ 83; 2; 58; 28; 600

Classroom Data (page 101)

- ● 8 students
- ■ 12 students
- ▲ 4 cans; 12 cans

Measure It (page 103)

Answers will vary as students will collect different data.

Keeping Track (page 105)

- ● 3 students
- ■ 27 students
- ▲ 48 students

Make It Yourself (page 107)

- ● 9 square units; any 3 × 3 square
- ■ 8 square units; there are multiple possibilities, for example a 4 × 4 square
- ▲ 9 square units; there are multiple possibilities, for example:

Which One? (page 109)

- ● D
- ■ A
- ▲ Madelyn drew picture B and Ben drew picture D.

Hobbies (page 111)

- ● 12 square inches; 24 square inches
- ■ 72 square inches; $A = 8 \times 9$ or $A = 8 \times 4 + 8 \times 5$ or $A = 8 (5 + 4)$
- ▲ Problems will vary.

All Around the Garden (page 113)

- ● 40 ft.
- ■ 4 ft.
- ▲ 28 ft.

Many Measures (page 115)

- ● 1 ft. × 12 ft.; 2 ft. × 6 ft.; 3 ft. × 4 ft. (in any order); The one with the perimeter of 16 ft. is 2 ft. × 6 ft.
- ■ 1 ft. × 18 ft.; 2 ft. × 9 ft.; 3 ft. × 6 ft. (in any order); The one with the shortest perimeter is 3 ft. × 6 ft.; 18 ft.
- ▲ The perimeters could be 50 ft., 28 ft., 22 ft., or 20 ft.; your rectangle's lengths are 4 ft. and 6 ft.; The lengths of your friend's rectangle are 2 ft. and 8 ft.

Answer Key *(cont.)*

Tangram Shapes (page 117)

(Note that the specific order of the pieces may vary.)

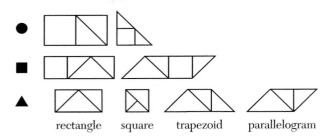

rectangle   square   trapezoid   parallelogram

Tell Me More (page 119)

● Possible answers include: I have four sides, I have four angles, the lengths of my sides are equal, or the measures of my angles are the same; checkerboards, floor tiles, or table tops

■ Possible answers include: I have four sides, opposite sides have the same length, the measures of my angles are the same; faces of boxes, walls, doors, windows, or picture frames; yes

▲ Possible answers include: I have four sides, opposite sides have the same length, two pairs of angles are the same; kites, baseball diamonds, some parking lot spaces, or pieces in some quilts; yes

Name the Shape (page 121)

● rectangle

■ rhombus or parallelogram or quadrilateral; any of the previous names not chosen

▲ square or rectangle or parallelogram or quadrilateral; the two previous names not chosen

Table Shapes (page 123)

● 14 chairs

■ 4 tables

▲ 7 pentagon tables and 4 hexagon tables

Shape Sentences (page 125)

● Possible correct answers include: rectangles, quadrilaterals, or parallelograms

■ Possible correct answers include: parallelograms, quadrilaterals, squares, or rectangles; pentagons or triangles

▲ Possible correct answers include: quadrilaterals; parallelograms; triangles, squares, pentagons, or hexagons

They Belong Together (page 127)

● C

■ D

▲ B

Draw Me (page 129)

Drawings will vary but should be checked against problem statements.

Parts of Shapes (page 131)

● 4 small squares; $\frac{1}{4}$

■ 5 small squares; $\frac{1}{6}$

▲ $\frac{1}{8}$

References Cited

Bright, G. W., and J. M. Joyner. 2005. *Dynamic classroom assessment: Linking mathematical understanding to instruction.* Chicago, IL: ETA Cuisenaire.

Brown, S. I., and M. I. Walter. 2005. *The art of problem posing.* Mahwah, NJ: Lawrence Earlbaum.

Cai, J. 2010. Helping elementary students become successful mathematical problem solvers. In *Teaching and learning mathematics: Translating research for elementary school teachers*, ed. D. V. Lambdin and F. K. Lester, Jr., 9–13. Reston, VA: NCTM.

D'Ambrosio, B. 2003. Teaching mathematics through problem solving: A historical perspective. In *Teaching mathematics through problem solving: Prekindergarten– Grade 6*, ed. F. K. Lester, Jr. and R. I. Charles, 37–50. Reston, VA: NCTM.

Goldenberg, E. P., N. Shteingold, and N. Feurzeig. 2003. Mathematical habits of mind for young children. In *Teaching mathematics through problem solving: Prekindergarten– Grade 6*, ed. F. K. Lester, Jr. and R. I. Charles, 51–61. Reston, VA: NCTM.

Michaels, S., C. O'Connor, and L. B. Resnick. 2008. Deliberative discourse idealized and realized: Accountable talk in the classroom and in civil life. *Studies in philosophy and education* 27 (4): 283–297.

National Center for Educational Statistics. 2010. Highlights from PISA 2009: Performance of U.S. 15-year-old students in reading, mathematics, and science literacy in an international context. http://nces.ed.gov/pubsearch/pubsinfo.asp?pubid=2011004

National Governors Association Center for Best Practices and Council of Chief State School Officers. 2010. Common core state standards. http://www.corestandards.org/the-standards.

National Mathematics Advisory Panel. 2008. *Foundations for success: The final report of the National Mathematics Advisory Panel.* Washington, DC: U.S. Department of Education.

Polya, G. 1945. *How to solve it: A new aspect of mathematical method.* Princeton, NJ: Princeton University Press.

Sylwester, R. 2003. *A biological brain in a cultural classroom.* Thousand Oaks, CA: Corwin Press.

Tomlinson, C. A. 2003. *Fulfilling the promise of the differentiated classroom: Strategies and tools for responsive teaching.* Alexandria, VA: ASCD.

Vygotsky, L. 1986. *Thought and language.* Cambridge, MA: MIT Press.

Contents of the Teacher Resource CD

Teacher Resources

Page	Resource	Filename
27–31	Common Core State Standards Correlation	ccss.pdf
N/A	NCTM Standards Correlation	nctm.pdf
N/A	TESOL Standards Correlation	tesol.pdf
N/A	McREL Standards Correlation	mcrel.pdf
132	Student Response Form	studentresponse.pdf
133	Individual Observation Form	individualobs.pdf
134	Group Observation Form	groupobs.pdf
135	Record-Keeping Chart	recordkeeping.pdf
N/A	Exit Card Template	exitcard.pdf

Lesson Resource Pages

Page	Lesson	Filename
33	Snack Time	snacktime.pdf
35	Floor Tiles	floortiles.pdf
37	Equal Groups	equalgroups.pdf
39	Boxes of Cupcakes	boxescupcakes.pdf
41	Pattern Questions	patternquestions.pdf
43	First Names	firstnames.pdf
45	Pose a Problem	poseproblem.pdf
47	Finish the Steps	finishsteps.pdf
49	Pattern Hunt	patternhunt.pdf
51	Boxes and Boxes	boxesboxes.pdf
53	Figure It	figureit.pdf
55	At the Fair	atfair.pdf
57	Yard Sale	yardsale.pdf
59	What's Going On?	whatsgoingon.pdf
61	Number Models	numbermodels.pdf
63	Buildopoly	buildopoly.pdf
65	Toy Store	toystore.pdf
67	Some Sums	somesums.pdf
69	Animals Facts	animalfacts.pdf
71	Family Trips	familytrips.pdf
73	Make It True	maketrue.pdf
75	Town Races	townraces.pdf

Contents of the Teacher Resource CD (cont.)

Lesson Resource Pages (cont.)

Page	Lesson	Filename
77	Make It Match	makematch.pdf
79	On the Number Line	numberline.pdf
81	On the Trail	ontrail.pdf
83	Standing in Line	standingline.pdf
85	Who Is Where?	whowhere.pdf
87	How Much Money?	howmuchmoney.pdf
89	What Is the Number?	whatnumber.pdf
91	Saturday Mornings	saturday.pdf
93	Pose the Question	posequestion.pdf
95	Balance It	balanceit.pdf
97	What Does It Hold?	whathold.pdf
99	Moving Along	moving.pdf
101	Classroom Data	classroom.pdf
103	Measure It	measureit.pdf
105	Keeping Track	keeptrack.pdf
107	Make It Yourself	makeyourself.pdf
109	Which One?	whichone.pdf
111	Hobbies	hobbies.pdf
113	All Around the Garden	aroundgarden.pdf
115	Many Measures	manymeasures.pdf
117	Tangram Shapes	tangram.pdf
119	Tell Me More	tellmemore.pdf
121	Name the Shape	nameshape.pdf
123	Table Shapes	tableshapes.pdf
125	Shape Sentences	shapesentences.pdf
127	They Belong Together	theybelong.pdf
129	Draw Me	drawme.pdf
131	Parts of Shapes	partsshapes.pdf

Contents of the Teacher Resource CD *(cont.)*

Additional Lesson Resources

Page	Resource	Filename
36	Crayon Box Picture	crayonbox.pdf
38	Cupcake Picture	cupcakepicture.pdf
56	Sale Picture	salepicture.pdf
60, 70	Number Line 0–100	numberline100.pdf
60, 70	Number Line 0–450	numberline450.pdf
88	Fraction Template	fractiontemplate.pdf
100	Sports Picture Graph	sportsgraph.pdf
104	Bar Graph Template	bargraph.pdf
120	Shape Matrix	shapematrix.pdf